The Biology of the Honey Bee

The Biology of the Honey Bee

Mark L. Winston

Harvard University Press
Cambridge, Massachusetts
London, England

Printed in the United States of America
FOURTH PRINTING, 1994

First Harvard University Press paperback edition, 1991

Library of Congress Cataloging-in-Publication Data

Winston, Mark L.
The biology of the honey bee.

Bibliography: p.
Includes index.
1. Honeybee. 2. Bees. I. Title.
QL568.A6W56 1987 595.79'9 86-31940
ISBN 0-674-07408-4 (alk. paper) (cloth)
ISBN 0-674-07409-2 (paper)

For Sue and Devora, with love

Preface

The objective of this book is to provide an in-depth introduction to the biology and social behavior of a single insect species, the honey bee. Because of its intrinsically interesting nature and economic value, the honey bee has been intensively studied from every scientific perspective, and a vast literature exists on almost every aspect of its existence. In this book I summarize what I consider the basic aspects of honey bee biology and provide references to that literature. I have tried to capture the flavor and richness of the many approaches which have been taken in studying the honey bee, but my own perspective and training is in entomology and behavioral ecology, and this book naturally reflects my own biases and background.

I would like to acknowledge here my teachers and colleagues who have had particular impact on my thinking about science and bees, especially Lynn Margulis, Charles Michener, Gard Otis, Keith Slessor, and Orley Taylor; their insights and influence pervade these pages. I am grateful to all those who read parts of this book and provided useful criticism: Cam Jay, Steve Kolmes, Charles Michener, Gard Otis, Gene Robinson, Tom Seeley, Keith Slessor, and especially Susan Katz. Some background research was done by Cynthia Scott and Mike Smirle, and the herculean task of organizing the references was ably conducted by Les Willis; I thank them for their assistance. I also thank the students in my beekeeping, bee biology, and social insects classes through the years, and hope that this book stimulates them to continue to ask questions.

I owe a deep debt of gratitude to Elizabeth Carefoot and the Instructional Media Center at Simon Fraser University, for the illustrations in this book and many other things. The illustrations are original drawings or were redrawn from other sources which are cited individually in the figure legends. I am grateful to the following publishers, journals, and authors for permission to redraw illustrations for which they hold the copyright: M. D. Allen; Bailliere Tindall (*Animal Behaviour*); Cornell University Press; H. A. Dade; Dadant and Sons; D. J. C. Fletcher; W. H. Freeman and Co. (*Scientific American*);

Harcourt, Brace, Jovanovich; Harvard University Press; International Bee Research Association; V. Lacher; Macmillan Journals (*Nature*); Masson S.A. (*Insectes Sociaux*); H. Martin; National Research Council of Canada (*Canadian Journal of Zoology*); Princeton University Press; Charles Scribner's; T. D. Seeley; and Springer-Verlag (*Behavioural Ecology and Sociobiology, Oecologia*). I am also grateful to the editors at Harvard University Press for their assistance at all stages of the writing and preparation of this book, particularly Elizabeth Hurwit and Angela von der Lippe.

Finally, I would like to acknowledge the financial assistance of the following agencies, which have provided substantial research support and in many ways made this book possible: the U.S. Department of Agriculture and National Science Foundation, the Natural Sciences and Engineering Research Council of Canada, the British Columbia Science Council, and Simon Fraser University.

Contents

The Biology of the Honey Bee

1

Introduction

The honey bee exhibits a combination of individual traits and social cooperation which is unparalleled in the animal kingdom. The multiple levels at which the honey bee expresses adaptations to its world provide one of the richest sources for study and knowledge among all organisms, made even more enriching by the economic benefits the honey bee provides. The honey bee can be, and has been, studied from perspectives as divergent as those of a beekeeper and a molecular biologist, a behavioral ecologist and a primitive honey hunter, a student of social behavior and a doctor interested in allergic reactions, and all have contributed immensely to our understanding of this most well-studied insect.

A quick glimpse inside the nest makes it readily apparent why honey bees have fascinated us from the earliest days of scientific observation (Fig. 1.1). The infrastructure of the nest, the exquisitely uniform and functional comb, is composed of beeswax produced by the workers and constructed into a repeating series of almost perfect hexagonal cells. The comb provides the substrate for interactions between colony members and is used for almost everything imaginable, from larval nursery to pantry to message center. As the stage for colony activities, it provides us with a rich arena for observing the individual and social behaviors that are at the heart of honey bee society.

At the individual level, honey bees have not one but three types of colony members: queens, drones, and workers, each with their own specializations and place in honey bee society. The aptly named queen reigns over the nest, surrounded by attendants and fed the rich food she requires to perform her few but crucial tasks in the colony. Her slim lines hide the huge ovaries which make her an extraordinary egg-laying machine, capable of laying thousands of eggs a day, and her calm behavior masks her powerful pheromones, chemical signals to recipient workers which control many of their behaviors and provide part of the social glue which holds honey bee life together.

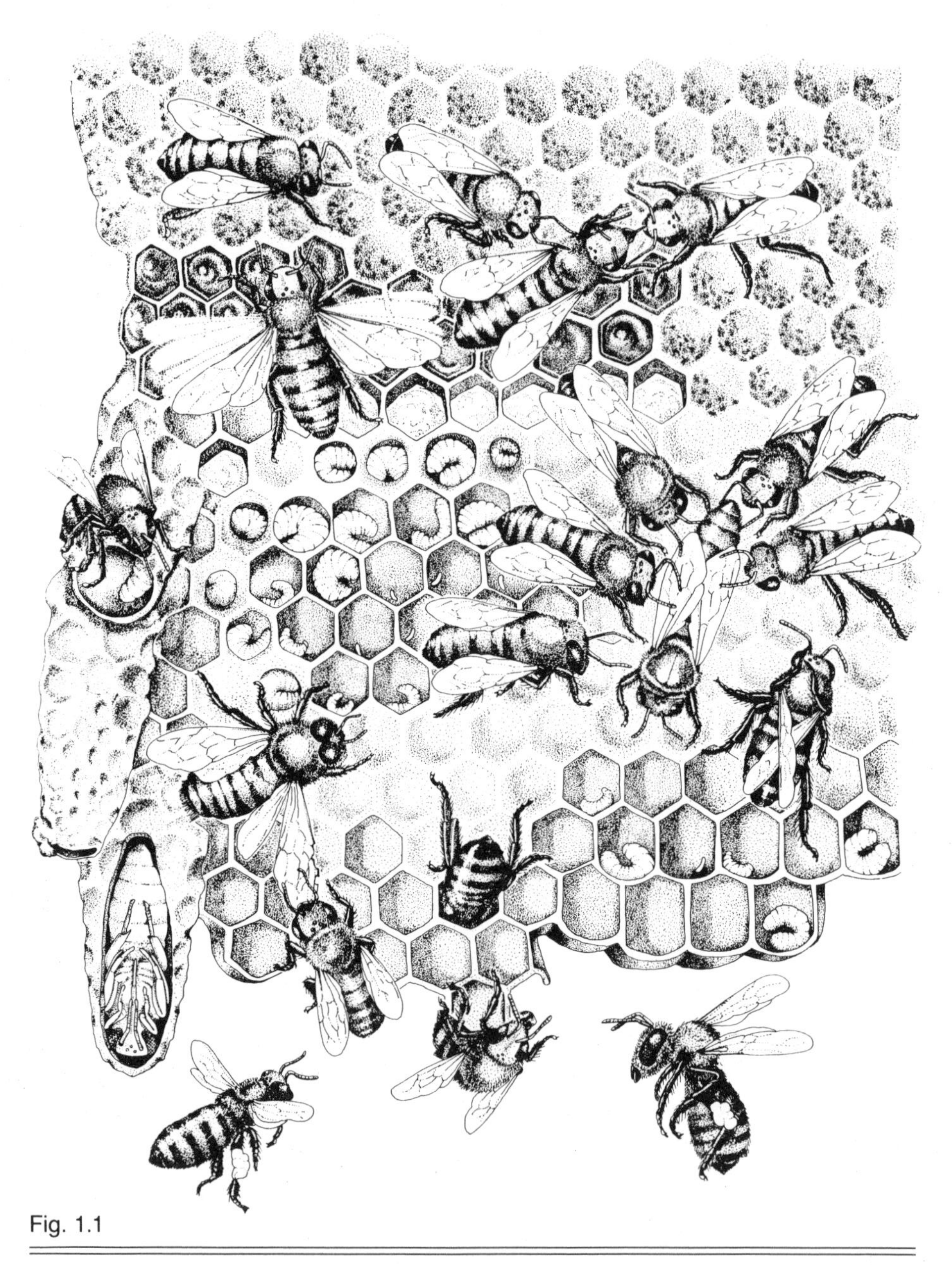

Fig. 1.1

A typical scene from a honey bee colony during the summer. Cells in the top section of the comb contain capped honey and uncapped nectar and pollen, while the middle and bottom cells contain eggs, larvae, and capped brood. The elongated cells on the left side of the comb are queen cells with various stages of developing queen brood inside; the bottom queen cell is cutaway to show a queen almost ready to emerge. The adult workers are, in clockwise order from the top left: resting, fanning to evaporate water from nectar, exchanging food, attending the queen in a retinue surrounding her, grooming, returning to the nest with nectar and pollen, building the comb, inspecting cells, driving a drone out of the nest, and capping a queen cell. (Based on Wilson, 1971.)

The other sexual members of this society, the drones, are tended and fed by the workers, although they perform only one function, the all-important one of mating the queen, after which they die. With their large eyes, flight muscles, and powerful mating urge, the drones are magnificently constructed for this one task.

The workers perform endless and diverse tasks in the nest, sometimes dying from stinging their colony's opponents, and rarely reproducing. At any one time workers may be found walking on the comb surface, perhaps tending the brood, cleaning debris from the nest, capping cells, ripening or storing honey, organizing pollen for storage near the brood, feeding or grooming the queen, or any one of a myriad other activities.

Not only have the complex individual behaviors manifested by these three castes prompted our investigations into honey bee life, but their sociality and adaptability has fueled the study of this insect to an extent not found for many other organisms on earth. The social nature of this organism, its tremendous ability to regulate its functions as a colony of individuals according to events within and outside the nest, provides the key to its success and makes the contribution of individuals much greater than the sum of their individual behaviors. We are further motivated to study this insect by an economic stake, since the products of the hive are crucially important to our agricultural systems. Not only do honey bees provide us with honey, wax, propolis, royal jelly, and pollen, but they also pollinate a good portion of our crops, including such diverse agricultural plants as fruit trees, oilseeds, small berries, and forage crops.

The study of honey bees is not new; extant cave paintings dating back many thousands of years show honey bee nests in marvelous detail. Certainly these are the forerunners of the thousands of articles written in the scientific and popular journals of contemporary science. It is an intangible quality of honey bee society that draws us to its study, an almost mystical presence which pervades their nests and can only be partly dissected, analyzed, and reported on by its observers. This book is an attempt to provide at least a glimpse into the world of this queen of insects.

2

The Origin and Evolutionary History of Bees

Bees are essentially wasps which abandoned predation in favor of provisioning nests with nectar and pollen. Most of the aculeate (stinging) wasps which are related to bees prey on other insects or spiders for larval food. However, adults frequently feed on nectar from flowers, and their mouthparts are often well adapted for sucking or lapping. Bees are thought to have evolved from a wasp ancestor, probably a sphecid, with mouthparts capable of ingesting nectar, which began collecting pollen to feed to their brood instead of killing prey. While bees have diverged from wasps in many characteristics (Michener, 1974), the most distinctive morphological differences involve specializations for pollen collection. All bees have at least a few plumose hairs and broadened hind legs, both adaptations for gathering pollen and transporting it back to the nest. Because of their distinctive pollen-collecting structures and habits, bees are classified in their own superfamily, Apoidea—order: Hymenoptera—(Culliney, 1983), although Michener (1974) has proposed returning to an older system which included the sphecoid wasps in the same superfamily as the bees.

The earliest bees may have arisen in the xeric interior of the paleocontinent Gondwana, which was probably the area of origin for flowering angiosperm plants (Raven and Axelrod, 1974). Although the fossil record of bees is far from complete, they are thought to have diverged from the sphecoid wasps during the middle Cretaceous period, about 100 million years ago (Michener, 1974), coincidental with the appearance of angiosperms as the dominant vegetation. The earliest known fossil bees are from the Eocene period 40 million years ago (Manning, 1952; Kelner-Pillault, 1969; Zeuner and Manning, 1976), but since these specimens were already highly specialized it is clear that bees arose much earlier. At any rate, the evolution and divergence of bees has been closely linked to that of the angiosperm plants, with the plants evolving flowers with odors, shapes, colors, and excess nectar and pollen food rewards to attract the bees, and the bees in turn providing a mechanism to transfer pollen between plants. The coevolution of these two groups has been one of the dominant themes of recent evolutionary history.

There are currently 10 or 11 families of bees (Michener, 1974; Michener and Greenberg, 1980), with approximately 700 genera (Malyshev, 1968) and 20,000 living species (Michener, 1969). These can be divided into two major groups, the more primitive short-tongued bees and the more advanced long-tongued bees (Fig. 2.1). The short-tongued bees probably utilized the shallow flowers characteristic of the earliest angiosperms, but some bees evolved longer mouthparts as many of the angiosperms developed longer, tubular flowers. These adaptations allowed the long-tongued bees to take advantage of the increasing complexity of advanced angiosperm flowers.

The honey bee *Apis mellifera* L. was one of these long-tongued bee species. Its scientific name means "honey-bearing or honey-producing bee," and refers to the bees' habit of collecting nectar and producing from it copious amounts of honey to allow colonies to survive dearth periods.

Honey bees are classified in the family Apidae, and their close relatives include the orchid bees (Euglossini), the bumble bees (Bombini), and the stingless bees (Meliponinae) (Winston and Michener, 1977; Kimsey, 1984) (Fig. 2.2). All of the Apidae are characterized by the presence of a corbicula or pollen basket on the outer surface of

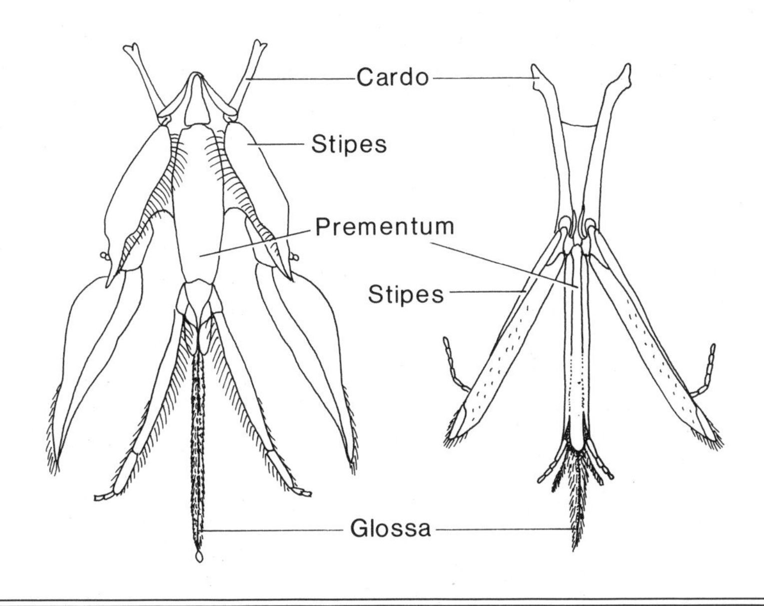

Fig. 2.1

Mouthparts of representative long-tongued (*left*) and short-tongued (*right*) bees. (Redrawn from Michener, 1974.)

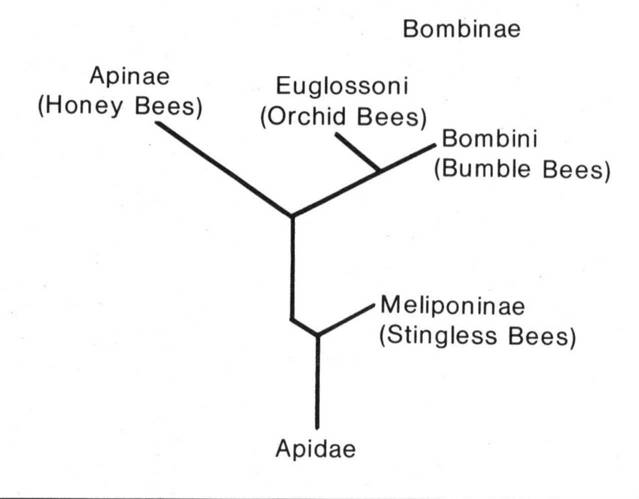

Fig. 2.2

Taxonomic relationships between bees in the family Apidae.

each hind tibia, at least in workers, and this structure is used to carry pollen and nest-building materials. Pollen is generally pushed up into the corbicula from the distal end of the tibia, and this loading mechanism is also a distinctive feature of the apids (Winston and Michener, 1977; Michener, Winston, and Jander, 1978). All of the Apidae show some degree of social behavior, and the Meliponinae and Apinae have the most elaborate social behavior of all the bees. Reviews of the natural history and biology of stingless bees and orchid bees can be found in Michener (1974), while Free and Butler (1959), Alford (1975), and Heinrich (1979a) have written accounts of bumble bee societies.

Honey Bee Origins

The modern honey bees (Apidae: Apini) are all classified in only one genus, *Apis*, which includes five species: the common honey bee *A. mellifera*, the giant honey bees *A. dorsata* and *A. laboriosa*, the Indian honey bee *A. cerana*, and the dwarf honey bee *A. florea*. The earliest Apini fossil specimens have been found in Baltic Amber from Eocene layers approximately 40 million years old, and these ancient but extinct apines have been classified in their own genus *Electrapis* (Manning, 1960; Zeuner and Manning, 1976; Culliney, 1983). A fossilized honey bee comb has recently been discovered in Malaysia dating from the late Tertiary or early Quaternary period, suggesting an earlier origin of the genus (Stauffer, 1979). This finding and the fact that fossil honey bee specimens are generally found with individuals grouped together suggest early evolution of social behavior in the Apini. Rapid evolution during the next 10 million years is shown by specimens found from the Oligocene, when considerable change occurred in external morphology. Also, comparative biochemical studies of extant bees have indicated a greater degree of amino acid substitution in *A. mellifera* compared to other bees, and therefore a more rapid protein evolutionary rate in the honey bee lineage than for other bees (Carlson and Brosemer, 1971, 1973). On the basis of morphological evidence, however, there has been relatively little change in honey bees during the last 30 million years (Culliney, 1983), and the physical resemblance of fossil forms to modern worker bees suggests that complex social behavior had already developed by the Miocene, 27 million years ago.

The natural geographic distribution of the genus *Apis* shows its greatest species diversity in India and adjacent regions, with all of the species except *A. mellifera* found there. Therefore, these regions probably constitute the area of origin and early evolution of the Apini (Doediker, Thakar, and Shaw, 1959; Michener, 1974; Doediker, 1978). *A. mellifera* is thought to have originated in the African tropics or subtropics during the Tertiary period, migrating to western Asia and colder European climates somewhat later. Until modern times *Apis* was not found anywhere in the western hemisphere, Australia, or the Pacific except for some of the continental islands such as Japan, Formosa, the Philippines, and Indonesia (Michener, 1974). But movement of bees by European settlers for beekeeping has resulted in *A. mellifera* now having worldwide distribution and some of the other species being more widespread in Asia.

The dwarf honey bee *A. florea* appears to maintain several ancestral characteristics of the genus *Apis* and is probably the closest living descendant of the earliest honey bees. Workers are small, approximately 7 mm in length, and colonies construct a single comb supported from branches, frequently at sites surrounded by dense vegetation (Seeley, Seeley, and Akratanakul, 1982). Their communicative dances occur on a horizontal platform built on the top of the comb, and thus direction to flowers is indicated by straight runs toward the food source. Colonies tend to be small, less than 5000 individuals, and the workers are relatively docile (Michener, 1974).

Two other closely related species of honey bees also construct nests consisting of a single comb in the open, the giant honey bees *A. dorsata* and *A. laboriosa*. These are large, feisty bees 17–19 mm in length, with 20,000 or more workers in a colony. Their nests are constructed high in trees or suspended from open cliff faces, and nests do not need to be concealed because of the worker's aggressive nature. Nests are also frequently aggregated, and colonies may migrate up and down mountains to take advantage of seasonal nectar sources. Communicative dances are more advanced than for *A. florea* since they occur on the vertical comb face, and the direction to flowers has to be translated by the workers from the vertical dance angle to the direction of the sun (Michener, 1974). *A. laboriosa* is the larger of the two species, and its large size, dark color, and long hair coat are probably adaptations for its high-altitude Himalayan habitat (Sakagami, Matsumura, and Ito, 1980).

The other two honey bee species, *A. cerana* and *A. mellifera,* are medium sized bees (10–11 mm) which generally build multiple comb nests inside cavities. Colonies of *A. cerana* are relatively small, 6000–7000 workers (Seeley, Seeley, and Akratanakul, 1982), but *A. mellifera* colonies can reach sizes of 100,000 or more individuals. These species are so similar in morphology and behavior that they have frequently been considered distant races of the same species. However, Ruttner and Maul (1983) recently demonstrated that, although *cerana* and *mellifera* queens and drones attempt to mate with each other, no offspring result, and instrumental insemination of *mellifera* and *cerana* queens with heterospecific semen revealed that hybrid fertilized eggs would cease development at the blastula stage. These results indicate that *mellifera* and *cerana* are indeed separate species, although closely related. The remainder of this book deals primarily with the honey bee *Apis mellifera,* and the term "honey bees" is used to refer only to this species, except where noted.

Races of *A. Mellifera*

The natural habitat of the honey bee *A. mellifera* extends from the southern tip of Africa through savannah, rain forest, desert, and the mild climate of the Mediterranean before reaching the limit of its range in northern Europe and southern Scandinavia. With such a variety of habitats, climatic conditions, and floras, it is not surprising to find numerous subspecies (races) of honey bees, each with distinctive characteristics adapted to each region (Louveaux, 1966) (Fig. 2.3). Still, recognition of valid races has been difficult for a number of reasons. Most important has been the movement of honey bees all over the world for beekeeping, which has changed the natural range of each race and resulted in considerable hybridization of races. Selection by beekeepers for characteristics useful in management may have altered the natural genotype of races as well, particularly in areas of intense beekeeping where many feral colonies are descended from swarms which have escaped from hives. Another difficulty may be that scientists and beekeepers don't always use the same criteria for determining what a "race" is. Scientists tend to use morphometric measurements of such characteristics as wing veins, mouthpart and

Fig. 2.3

The origins of the various *Apis mellifera* subspecies, and the westernmost extension of the range of *Apis cerana*.

antennal length, and the size of certain body parts (Ruttner, 1975a; Daly and Balling, 1978; Ruttner, Tassencourt, and Louveaux, 1978), whereas beekeepers prefer characteristics like color and behavioral traits, such as tendency to swarm, good honey production, and gentleness. Finally, even within a single race there can be tremendous variation, and where to divide races and determining what is "typical" for a race have always been somewhat subjective.

Some general conclusions have emerged concerning the characteristics and places of origin for many honey bee races, and these have been summarized by Ruttner (1975b; Ruttner, Tassencourt, and Louveaux, 1978). He divides honey bee races into three distinct groups: European, Oriental (Near Eastern), and African. Little is known about the Oriental races, and studies of many African regions are based on few specimens. The European races have been relatively well studied, and there appears to be more general agreement on these than on the African races. The brief descriptions below are based on Ruttner's conclusions except where noted.

EUROPEAN RACES

Apis mellifera mellifera L. (German dark bees) originated in northern Europe and west-central Russia, probably extending down into the Iberian peninsula. They are large bees, although with relatively short tongues (5.7–6.4 mm), and their common name is derived from their brown-black color, with only a few lighter yellow spots on the abdomen. They tend to be nervous and aggressive but winter well in severe climates. Worker population expands slowly in the spring, and, although these bees were once popular for export around the world, their aggressive nature, poor spring and early summer performance, and difficulty in working flowers with long corollas such as clover have resulted in diminished use of *mellifera* for beekeeping.

Apis mellifera ligustica Spin. (Italian bees), which originated in Italy, has been the most popular honey bee for beekeeping throughout the world. Although somewhat smaller than *mellifera, ligustica* have relatively long tongues (6.3–6.6 mm) and abdomens with bright yellow bands. They tend to be docile, and colony populations build up quickly in the spring and remain strong throughout the summer. They overwinter with strong worker populations, although with high consumption of honey that causes some difficulties in northern latitudes. They also have a reputation as rapid comb builders and seem to initiate robbing of honey from other colonies more quickly than the other European races.

Apis mellifera carnica Pollman (Carniolan bees) originated in the southern Austrian Alps, northern Yugoslavia, and the Danube Valley. They are of a similar size to *ligustica,* but tend to be gray or brown in color. These bees have also been popular for beekeeping, particularly with hobbyists because of their gentle disposition. They overwinter in small colonies with low food consumption but develop quickly in the spring. They may not maintain this high population throughout the summer and seem to swarm more readily than the Italian bees. They are also slow to construct new comb.

Apis mellifera caucasica Gorb. (Caucasian bees) evolved in the high valleys of the central Caucasus. They appear similar to *carnica,* although perhaps with a more lead-gray color. Although their behavior is not as well known, they are considered gentle, slow to expand in the spring but capable of reaching adequate summer populations, and poor at overwintering because of susceptibility to the adult disease Nosema. They also use propolis extensively and are purported to have only a weak disposition to swarm.

There may be a number of other European races, which have either been insufficiently studied or grouped within one of the other European groups. The Macedonian bee *A. m. cecropia* Kiesw now appears to belong to the *carnica* race, but the position of the Russian steppe bee *A. m. acervorum* and the transcaucasian *A. m. remipes* are not as clear.

AFRICAN RACES

Apis mellifera intermissa v. Buttel-Reepen (Tellian bees) is a North African race, found north of the Sahara from Libya to Morocco. A small, dark bee, it reputedly is aggressive and swarms frequently, rearing over 100 queens in each swarming period. During droughts, over 80% of colonies may die; owing to intensive swarming colony numbers rebound when conditions improve (Louveaux, cited in Ruttner, 1975b).

Apis mellifera lamarckii Cockerell (Egyptian bees, formerly named *A. m. fasciata*) are found in northeast Africa, primarily in Egypt and the Sudan along the Nile Valley. Like *intermissa,* they rear numerous queens, with one colony recorded as rearing 368 queen cells and producing one small swarm with 30 queens. They appear to be more closely related to the bees of Central Africa, however, based on similarities of dance dialects between *lamarckii* and *adansonii* (von Frisch, 1967a).

Apis mellifera scutellata Lepeletier (East African bees), throughout much of their range, were considered to be *adansonii* (Smith, 1961) until Ruttner (1975b) proposed that these bees from the savannahs of central and equatorial East Africa and most of South Africa were actually a separate subspecies, *A. m. scutellata.* This has created some confusion, since the African bees introduced to Brazil in 1956 were thought to be *adansonii,* and all of the literature concerning these bees prior to the mid-seventies refers to them as *adansonii.* There is still some disagreement about whether *scutellata* and *adansonii* are indeed different subspecies, and also about which subspecies was introduced to Brazil. Since Ruttner's 1975 study is the most recent and complete taxonomic evaluation of African bees, I have adopted his classification, although the status of these subspecies is currently being reevaluated and additional evidence may result in further changes. *A. m. scutellata,* a small bee with a relatively short tongue, is highly aggressive, swarms and absconds frequently, and is able to nest in a broad range of sites from cavities to open nests.

Apis mellifera adansonii Latreille (West African bees) are found in West Africa and are conspicuously yellow in color. They appear to be similar to *scutellata* in many of their behaviors, but bees from this region have not been well studied.

Apis mellifera monticola Smith (Mountain bees) are of interest because of the high altitude at which they are found in Tanzania, from 1500 to 3100 m. It is a large, dark, gentle race, with longer hairs than the other African bees.

Apis mellifera capensis Escholtz (Cape bees) are only found at the tip of South Africa and are unique among *Apis mellifera* in the common occurrence of female-producing laying workers. They are morphologically similar to *scutellata,* but the degree of ovariole development and ability to lay parthenogenetic females regularly separates them from the *scutellata* group.

Other African races are found in limited areas of Africa. These may be morphometrically distinguishable from the other races, but only a few specimens of each have been examined, and their biology has not been studied well enough to reach firm conclusions about their taxonomic status. These subspecies include *A. m. major* Ruttner, *sahariensis* Baldensperger, *nubica* Ruttner, *littorea* Smith, *unicolor* Latreille, and *jemenetica* Ruttner (Dutton et al., 1981).

ORIENTAL RACES

A number of oriental races from Turkey west to Iran have been proposed, including *A. m. syriaca, anatolia,* and *meda,* which is similar to *ligustica* (Ruttner, Pourasghar, and Kauhausen, 1985). The relationships between these groups have not been studied. A thorough evaluation of the systematics of oriental honey bees might be important, since presumably transition forms between temperate and tropically evolved races might be found, possibly between *A. mellifera* and *A. cerana.*

NORTH AND SOUTH AMERICAN RACES

Although honey bees are not native to North or South America, both European and African races have been introduced in the last few hundred years. In North America racial lines of European origin have generally been maintained, although extensive matings between races and different selective criteria by queen breeders have undoubtedly modified some of the bee's original characteristics. For simplicity, the original racial designations will be used here. In South America the introduction of African bees in 1956 has resulted in the establishment and spread of *A. m. scutellata* throughout much of South and Central America. These bees will be referred to as "Africanized" to differentiate them from bees studied in Africa, but they appear to be morphologically, behaviorally, and ecologically almost identical to *scutellata,* and thus do not constitute a separate race.

3

Form and Function: Honey Bee Anatomy

There is perhaps no more elegant aspect of honey bee biology than the relationship between form and function in adult bees. A complex combination of parts integrated into a finely tuned organism, the honey bee is capable of performing a broad range of athletic, graceful, and purposeful tasks. A list of all the parts which make up a honey bee is impressive in length, but more striking are the ways in which bee structures are designed to carry out their functions. I have derived much of the following discussion from more detailed reviews of honey bee anatomy in Snodgrass (1956) and Dade (1977).

Overview of Worker Anatomy

A honey bee is made up of a series of hardened plates connected by membranes and covered in most regions by a dense pile of hairs (Fig. 3.1). This external skeleton provides protection from predators, prevents water loss, serves as a framework for internal muscle attachment, and allows rapid but precise movements by a complex arrangement of internal ridges which the muscles can contract against. The exoskeleton and internal parts are arranged in three regions, the head, thorax and abdomen, each of which has a number of segments. The head functions largely to ingest and partially digest food through the mouthparts and associated glands, and also serves as the major sensory region of the body through the eyes, antennae, and sensory hairs. The thorax is made up of three segments, each bearing one pair of legs; in addition, each of the posterior two thoracic segments has one pair of wings. Thus, the thorax is the principal locomotory region of a bee's body and contains powerful muscles for flight, walking, and specialized functions such as pollen collecting. The abdomen consists of seven visible segments and contains the bulk of the internal organ systems as well as a structure of great interest to beekeepers, the sting.

Fig. 3.1

The body plan of a honey bee worker, showing the three major body regions, the mouthparts extended, the three pairs of legs, and the two pairs of wings.

The Head

Viewed from the front and magnified under a microscope, the head of a worker bee can be a frightening sight; the large eyes and jaws seem particularly menacing. In fact, the head is a highly specialized and relatively harmless region of a bee, capable of sensitive perceptions of the environment and manipulation and ingestion of the two major constituents of bee food, nectar and pollen, each of which provides different anatomical challenges for handling (Fig. 3.2).

SENSORY STRUCTURES

Visual perception occurs through two different types of eyes, the ocelli and the compound eyes, both found on the facial region of the head. The ocelli are actually three eyes arranged in a triangular pattern, each eye consisting of a simple, dense lens, which is derived from a thickening of the head exoskeleton, and sensory retinal cells beneath the lens (Yanase and Kataoka, 1963). The ocelli can not focus or make images and appear to function solely to detect light intensity, possibly to regulate diurnal activity patterns or for orientation (Lindauer and Schricker, 1963; Schricker, 1965).

The two compound eyes take up a substantial part of the head and are complex visual organs capable of a wide range of photoreceptive functions. Each worker compound eye is made up of about 6900 hexagonal facets, each containing its own lens to receive light, a pigmented cone to concentrate and focus it, and sensory retinal cells for light perception (Fig. 3.3). Each facet responds independently to the incoming light waves, and groups of facets are specialized for perceiving the plane of light polarization (von Frisch, 1967a; Edrich and von Helversen, 1976; Zolotov and Frantsevich, 1973; Wehner and Strasser, 1985), pattern recognition (Wehner, 1972), color vision (Kaiser, 1972; Menzel, 1973), and head turning responses (Moore, Peni-

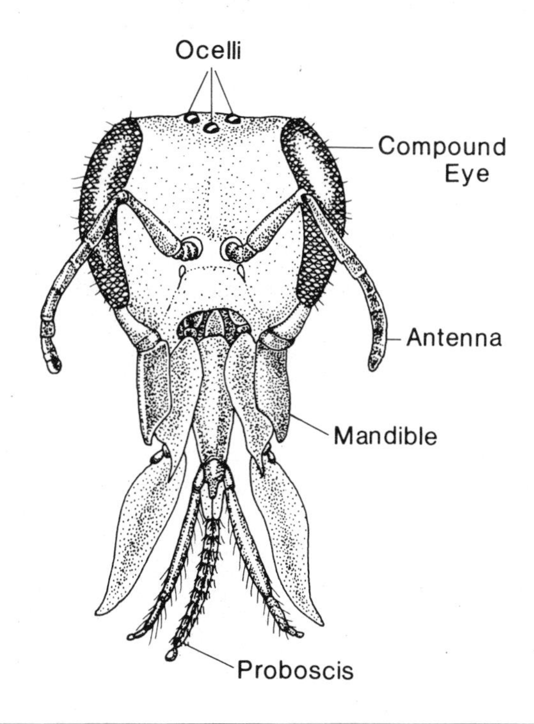

Fig. 3.2

The head of a worker bee, with the proboscis extended. (Redrawn from Snodgrass, 1956. Copyright © 1956 by Cornell University. Used by permission of Cornell University Press.)

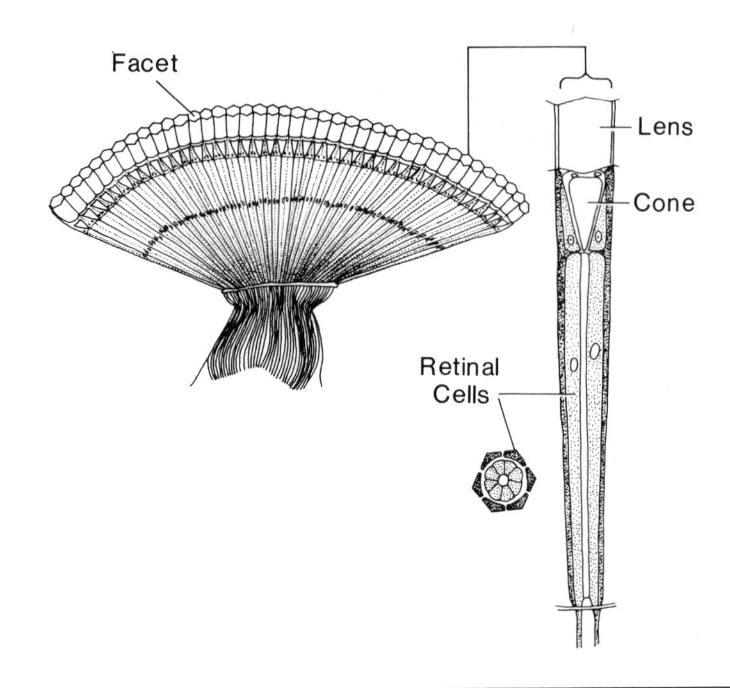

Fig. 3.3

Cross section through a worker's eye, showing some of the individual facets. One of the facets has been enlarged to show the structures which receive, concentrate, and perceive light. (Redrawn from Snodgrass, 1956. Copyright © 1956 by Cornell University. Used by permission of Cornell University Press.)

kas, and Rankin, 1981). Images are produced by the central nervous system integrating the signals from the individual facets into a mosaic image. Since each facet diverges angularly only about 1° from its neighbors (del Portillo, 1936; Michener, 1974), the mosaic pattern is particularly well adapted for detecting movement. In addition to visual motion perception, the compound eyes can perceive airflow using sensory hairs arranged at the junctions of the facets (Fig. 3.2). When these hairs are removed with a tiny scalpel, workers lose their ability to find their way to accustomed feeding sites under windy conditions, presumably because they can no longer compensate for wind speed during flight (Neese, 1965).

The antennae are the noses of the bee, each consisting of a ten-segment flagellum attached to a scape and pedicel at the base (Fig. 3.4). The role of bee antennae in smell was first demonstrated by von Frisch (reviewed in von Frisch, 1967a), who showed that workers could be trained to visit dishes which contained odors of natural flowers or essential oils. When the antennae were surgically removed, this olfactory discrimination ability was eliminated. Subsequent experiments showed that the olfactory acuity of bees is approximately equal that of man, although workers are 10 to 100 times more sensitive to wax, flower, and other odors biologically significant to bees (Ribbands, 1955; Schwarz, 1955; Fischer, 1957; Vareschi, 1971).

In addition to their acute olfactory sensitivity, bees also have what

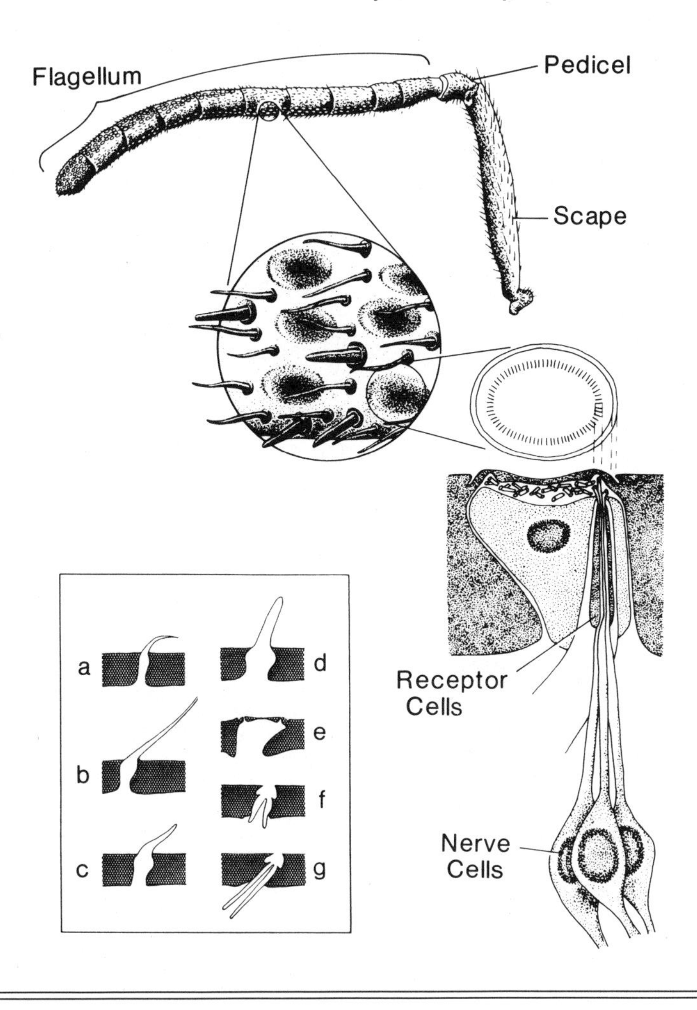

Fig. 3.4

The antenna of a worker bee, with one of the pore plates enlarged to reveal the odorant receptor apparatus. The insert shows the seven types of sensory structures found on antennae: (*a*) small thick-walled hair (sensillum trichodeum), (*b*) thick-walled peg (s. trichodeum), (*c*) slender thin-walled peg (s. trichodeum olfactorium), (*d*) large thin-walled peg (s. basiconicum), (*e*) pore plate or plate organ (s. placodeum), (*f*) pit organ (s. coeloconicum), and (*g*) pit organ (s. ampullaceum). (Nomenclature is from Lacher, 1964. Redrawn from von Frisch, 1967a, based on Lacher, 1964, and Snodgrass, 1956. Copyright © 1956 by Cornell University. Used by permission of Cornell University Press.)

Forel (1910) called a "topochemical olfactory sense." That is, bees can use their paired antennae to detect accurately the direction of an odor by comparing the intensity of odorant molecules perceived by each antenna. This sense was demonstrated by experiments in which food rewards were used to train worker bees to odors in a Y-shaped tube (Lindauer and Martin, 1963; Martin, 1964) (Fig. 3.5). When the antennae of trained bees were glued in a crossed position, the workers chose the wrong direction at the fork, indicating that the antennae were providing information concerning odorant location.

The actual sense organs consist of at least seven types of sensory structures, ranging from pits to plates to hairs (Slifer and Sekhon, 1961; Lacher, 1964; von Frisch, 1967a; Dietz and Humphreys, 1971) (Fig. 3.4). Only the pore plates are known to be olfactory, based on behavioral and electrophysiological responses to odors (Lacher and Schneider, 1963; Lacher, 1964), and there are approximately 3000 of them on a single worker antenna. These pore plates consist of rows of very fine pores through which odorant molecules can pass and be transported to receptor cells beneath the plate. The plates are located only on the eight most distal flagellar segments, and amputation of those segments eliminates worker orientation to food-associated odors (von Frisch, 1921; Ribbands, 1955). Some races of bees have more of these plates than others, but no link between plate density and sensory acuity has been established (Stort and Barelli, 1981).

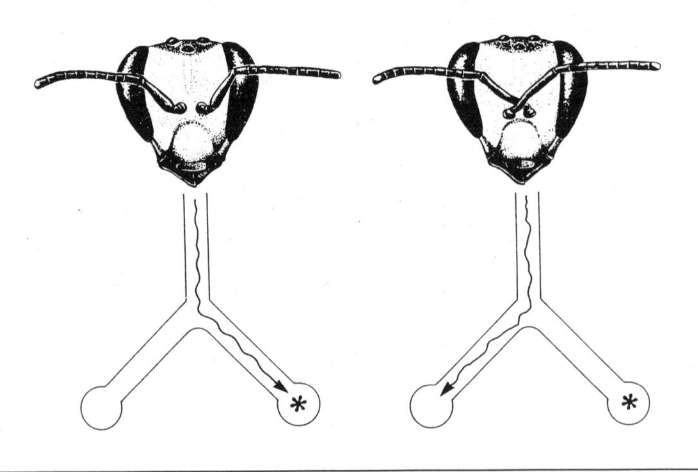

Fig. 3.5

Representation of the experiments revealing how workers use their paired antennae to detect the direction of an odor. Workers compare the intensity of odorants perceived by each antenna, which can be demonstrated by training workers to an odorant source (*), and then crossing their antennae, which causes workers to choose the wrong direction at the fork. (Redrawn from von Frisch, 1967a, based on Martin, 1964.)

The functions of the other sensilla types are not known, although there is good evidence that they are important for perception of carbon dioxide, humidity, taste, and possibly temperature. Using electrophysiological techniques, Lacher (1964) demonstrated that workers have a carbon dioxide receptor on the antennae, and Seeley (1974) showed that workers can detect differences in carbon dioxide levels of less than 1% and respond to high levels of carbon dioxide in the nest by fanning to increase air circulation. Moisture can also be detected by antennal receptors (Ribbands, 1955; Kuwabara and Takeda, 1956), with relative humidity differences as low as 5% being detectable by workers (Kiechle, 1961). Taste responses by antennae were demonstrated by touching antennae with sugar solutions and eliciting mouthpart responses (Minnich, 1932; Marshall, 1935a,b).

The antennae have one other structure of functional importance, the Johnston's organs. These are concentrations of sensory cells located internally in the pedicel of each antenna which are sensitive to minute changes in antennal position. Their function includes detection of airflow by the amount of antennal bending as a means of measuring flight speed (Heran, 1959).

MOUTHPARTS

The mouthparts of honey bees are classified as the chewing and lapping type, meaning that they can manipulate solid material as well as lap up liquids. They consist of the paired mandibles or jaws attached on the sides of the head and the proboscis or tongue, made up of the maxillae and the labium (Fig. 3.6).

The mandibles are strong, spoon-shaped jaws which are concave and ridged on the inner side (Michener and Fraser, 1978). Powerful muscles connect them to the head, and a channel fringed with hairs is found at the base of the duct leading from the mandibular glands. They have numerous functions in the workers, including ingesting pollen for food; cutting, shaping, and manipulating wax and propolis (plant resins) for nest construction; feeding brood food to larvae and nectar to the queen using the concave inner surface as a conduit from the mouth; dragging debris and dead bees out of the nest; grooming; and fighting.

The proboscis is a more complicated structure which has as its major function the ingestion of liquid materials, primarily nectar, honey, and water. The proboscis also functions in food exchange between workers as well as between workers and the queen, and workers and drones. It is also used to lick pheromones from the queen and exchange them with other workers. It is supported at the proximal end by two rods which connect it to the base of the mouth, and both the lateral maxillae and central labial structures are jointed so that the

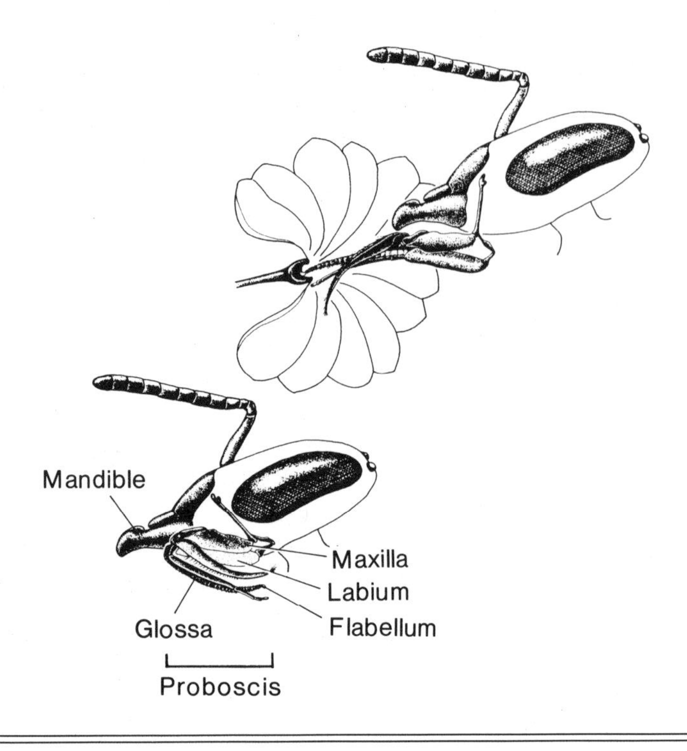

Fig. 3.6

Side view of worker mouthparts, including one mandible and the proboscis. When not in use, the proboscis is folded in a Z-shaped configuration into a cavity beneath the mouth, but can be fully extended to collect nectar, pollen, and water.

entire proboscis can be folded into the mouth cavity in a Z-shaped pattern when at rest. When fully extended, the galeae of the maxillae and the palps of the labium form a supporting tube around the tongue itself, the glossa (Winston, 1979a). The extended worker proboscis can be 5.3–7.2 mm long (Ruttner, Tassencourt, and Louveaux, 1978), depending on the race of bee, and this proboscidial length partly determines which flowers the workers can visit for nectar, since the proboscis must fit well into floral corollas in order to suck up nectar from the base of each flower.

The glossa itself is a densely hairy tongue which has hardened plates for strength, alternating with soft membranous areas for flexibility, and a long rod running from tip to base for additional support. There is a flabellum at the tip through which liquids are absorbed and transported up a narrow channel into the mouth (Michener and Brooks, 1984). At the base of the glossa is a group of muscles surrounding a hollow cavity, the cibarium, which is used to pump liquids. When the glossa is extended, these muscles as well as the paraglossae and other plates at the base form an airtight chamber. Movement of liquid is facilitated by back and forth glossal movement, capillary action, and pumping by the muscles of the cibarium which create suction for ingestion. The glossa is also important for pollen collection, since pollen grains frequently are trapped by the glossal hairs and then groomed from the proboscis by the forelegs (Michener, Winston, and Jander, 1978).

The Thorax

The bee thorax consists of three body segments, as for other insects, and the first abdominal segment is called a propodeum; incorporation of this segment into the thorax is a unique feature of most Hymenoptera. The three thoracic segments of bees in general are highly modified and specialized because of the combination of functions involving these segments: the wings are used for flight, and the legs perform a variety of functions including pollen collecting and grooming. Much of the shape and construction of the central thoracic parts can best be understood as adaptations for muscles which control movement, and a detailed review of thoracic sclerites and musculature can be found in Snodgrass (1956). Here, I concentrate on the structures and functions of the legs and wings.

LEGS

The six legs found on the thorax, one pair per thoracic segment, each have the same basic construction, although the workers' hind legs are highly modified to carry pollen and propolis. Each leg articulates with the main body of the thorax at the coxal segment, which allows for forward and backward leg motion (Fig. 3.7). The next leg segment, the trochanter, connects the basal coxa with the longer and more slender distal leg segments, the femur, tibia, and tarsus. The tarsus consists of five subsegments, the elongated basitarsus and four smaller tarsomeres. Finally, the tip of the leg consists of a terminal segment, the pretarsus, which includes the tarsal claws and associated pads (Fig. 3.8). These terminal structures are important in walking, since the claws and suction-creating pads grip the substrate and allow bees to walk on horizontal and vertical substrates, as well as to hang on to other bees in clusters. The tarsal claws are also used to some extent to manipulate wax for comb building.

The legs of workers perform their most refined functions in grooming and in packing pollen and propolis onto the hind leg for transportation back to the nest. The forelegs have hairy brushes on the enlarged basitarsus which are used to clean dust, pollen, and other foreign material from the head. The forelegs also carry the antenna

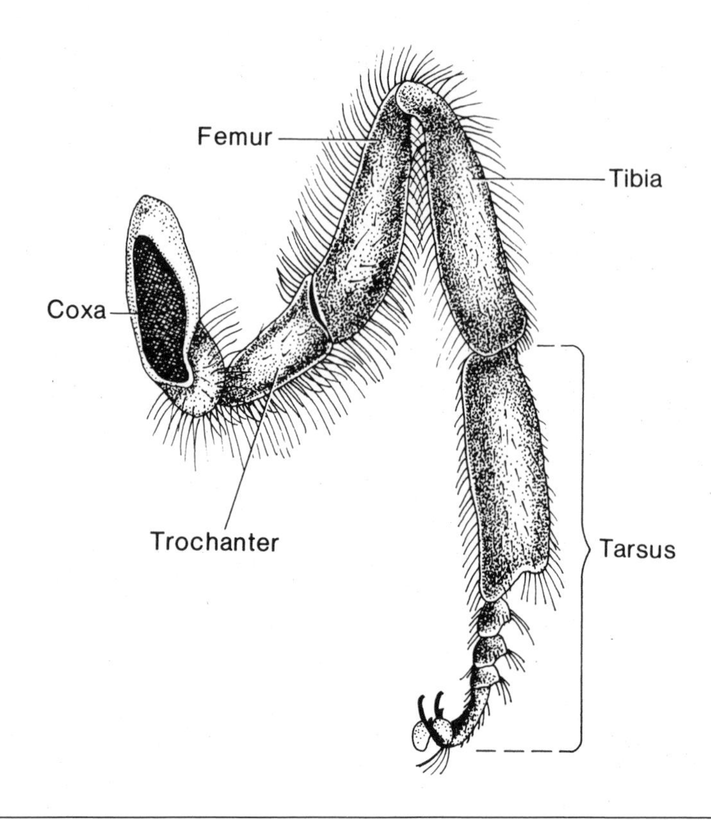

Fig. 3.7

Outer view of the middle leg of a worker bee. (Redrawn from Snodgrass, 1956. Copyright © 1956 by Cornell University. Used by permission of Cornell University Press.)

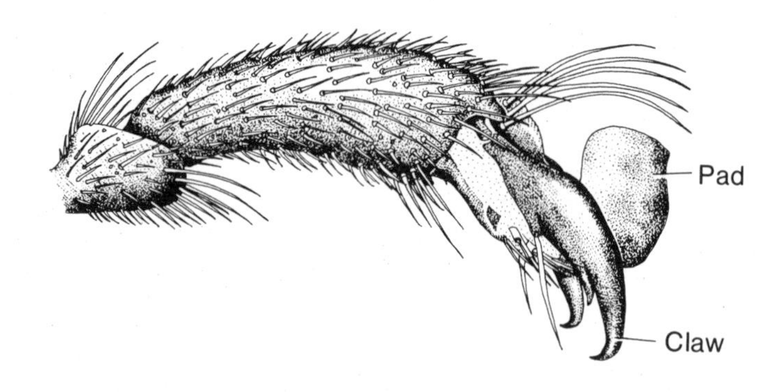

Fig. 3.8

A typical worker pretarsal leg segment, showing the tarsal pad and claws important for walking and manipulating substances such as wax and propolis. (Redrawn from Snodgrass, 1956. Copyright © 1956 by Cornell University. Used by permission of Cornell University Press.)

cleaner, a curved notch and associated spur through which the antennae can be pulled and brushed clean, at the junction of the tibia and basitarsus (Schönitzer and Renner, 1984) (Fig. 3.9). This structure undoubtedly is important in keeping the antenna clear of any material that might interfere with its sensory functions. The middle legs, although hairy, do not appear to be otherwise modified, and they are used to rid the thoracic hairs of dirt and pollen and to transfer material from the front to the hind legs. The middle legs do have a "blind spot" which they cannot reach to groom, the top of the middle thoracic segment, and workers returning from pollen-collecting visits to certain flowers can be seen with a stripe of brightly colored pollen on that region of the thorax.

The hind legs are highly modified for their pollen- and propolis-transporting functions (Fig. 3.10). The most prominent structure is the pollen basket, or corbicula, an expanded, slightly concave region on the outer surface of each tibia that is fringed by hairs on the edges and contains one central bristle upon which the pollen or propolis loads can be anchored (Hodges, 1967). Additional structures used only for pollen packing are found on the inner surface of the basitarsus, which has a regular series of stiff bristles called pollen combs as well as a flattened area at the base, the pollen press. The pollen-manipulating structures are completed by the rastellum or pollen rake, a stiff row of bristles on the inner edge of the tibia.

The corbiculae are thought to have originally evolved as baskets to carry sticky propolis back to the nest; the other hind leg modifications probably originated later for pollen collection (Winston and

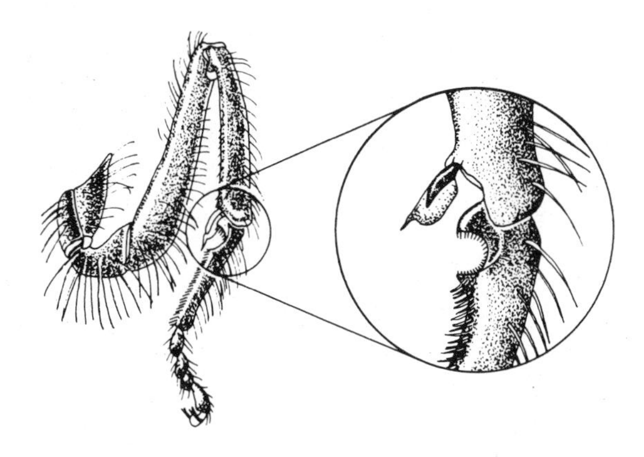

Fig. 3.9

The antenna cleaner found on the forelegs of workers. The antennae can be pulled through the notch and brush for grooming. (Redrawn from Snodgrass, 1956. Copyright © 1956 by Cornell University. Used by permission of Cornell University Press.)

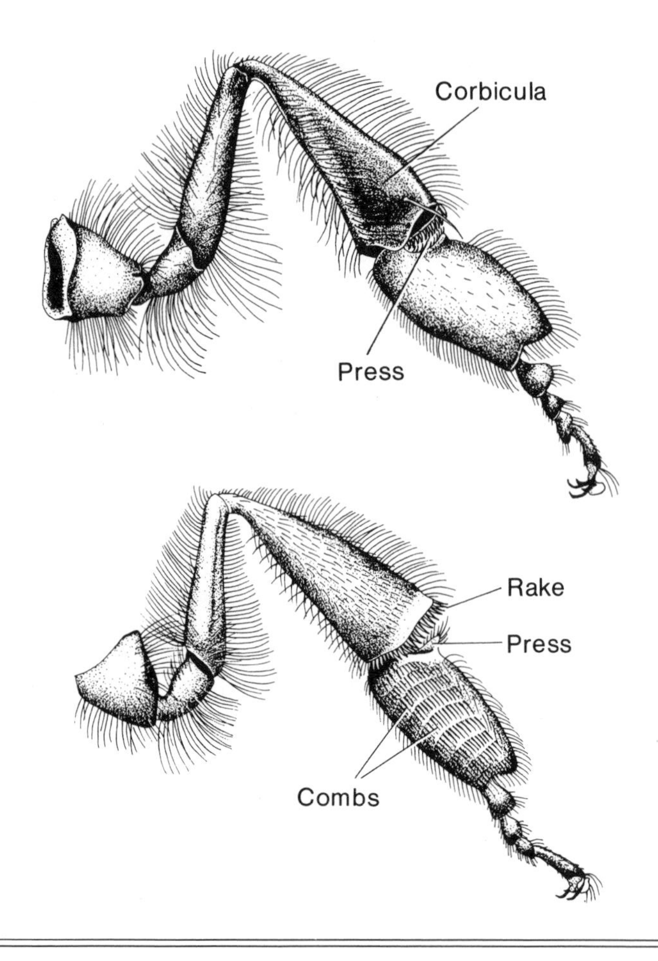

Fig. 3.10

Outer (*top*) and inner (*bottom*) views of a worker hind leg. The outer surface has the corbicula, or pollen basket, upon which pollen is transported, and the press, which pushes the pollen into the basket. On the inner surface are found the combs and rake, which manipulate pollen into the press for packing. (Redrawn from Snodgrass, 1956. Copyright © 1956 by Cornell University. Used by permission of Cornell University Press.)

Michener, 1977; Michener, Winston, and Jander, 1978). To obtain a load of propolis, workers first use their mandibles to bite off pieces of plant resins from buds and then pass the propolis on to the forelegs (Fig. 3.11). The load is then transferred to the inner surface of the basitarsus of the middle leg, on the same side of the worker's body. The middle leg is next used to press the propolis into the corbicula of the hind leg, again on the same side. Finally, the worker returns to the nest and the propolis is unloaded by the mandibles of another worker and pressed into an area of the nest needing plugging (Rösch,

1927; Meyer, 1956a; Jander, 1976; Michener, Winston, and Jander, 1978).

Pollen collection and packing in honey bees has been summarized by numerous authors (Hodges, 1952; Michener, Winston, and Jander, 1978; and other sources cited therein). Pollen is gathered from floral anthers by active movements of the legs and the proboscis scraping the anthers as well as by pollen which drops passively onto the body hairs of workers being groomed by the legs. The forelegs brush the proboscis, picking up pollen made sticky with regurgitated honey,

Fig. 3.11

Propolis collecting and packing techniques. Workers use their mandibles to scrape resin from plants, pass the propolis to the forelegs for transfer to the middle legs, and then press the propolis into the basket on the same side of the worker's body.

and also clean pollen from the head and the front of the thorax. The worker then takes to the air and hovers, transferring pollen from the forelegs and the posterior thoracic segments to the middle legs. The pollen now on the middle legs is passed to the pollen combs on the inner hind basitarsi by scraping the middle legs past the combs; the combs also scrape pollen from the abdomen.

The next transfer of pollen, from the inner basitarsal combs to the outer pollen basket, seems anatomically impossible, but honey bees have evolved an ingenious mechanism for pollen transfer and packing (Fig. 3.12). The pollen rake of the opposing hind leg scrapes the inner surface of each pollen comb, which results in the pollen being transferred from each pollen comb through the rastellum to the pollen press on the opposite leg. This can be observed as the hind legs rapidly rub against each other in flight. Finally, the pollen which has accumulated on each press is forced into the pollen basket by pumping the legs, and a sticky pollen pellet forms as more and more moistened pollen is pumped into the pollen basket. When the pollen-laden worker returns to the hive, the pollen pellets are removed from the basket by the middle legs and placed in cells, where they are packed for storage by other workers using their mandibles and forelegs to press the pollen into the cells.

WINGS

The wings of honey bees, as in all insects, are not true appendages like the legs but are thin outgrowths of the skeleton which have been substantially modified for flight (Fig. 3.13). The two pairs of wings are found on the posterior thoracic segments and articulate with the thorax in complex joints that allow a great range of movement. In the honey bee the front wings are larger than the hind wings, which can be attached to the front wings during flight by hooks called hamuli so that the two pairs of wings can beat in synchrony. This considerably reduces turbulence and drag during flight. The wings also contain veins which not only strengthen the thin wings but also carry blood, breathing tubes, and nerves to the wing extremities.

A worker's wings beat in flight at a rate of over 200 cycles/sec, which according to anatomical and aerodynamic analyses of flight is not possible. Thus there must be some special mechanisms in bees which sustain this high wingbeat frequency. One mechanism involves the type of nervous control over flight muscles; since the nervous system cannot operate quickly enough to keep up with the speed of muscle contractions to sustain flight, the thoracic muscles resonate. That is, they contract more than once for each nerve firing (Esch and Bastian, 1968; Esch, 1976; Bastian and Esch, 1970). Also,

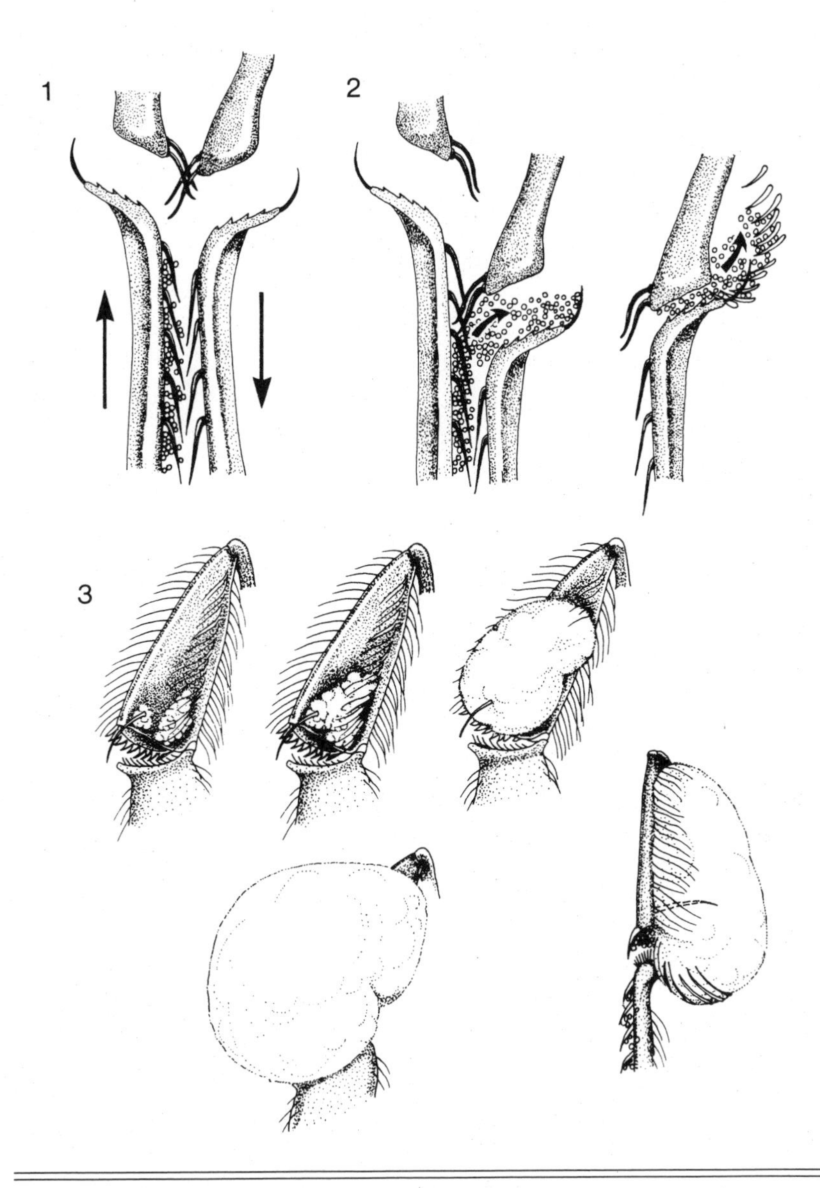

Fig. 3.12

Transfer of pollen from the inner surface of the hind legs to the outer surface of the opposing hind leg. (1) The pollen rake scrapes pollen from the opposing combs and deposits it on the press. (2) Pollen is then pushed into the corbicula by pumping the legs, and (3) transported back to the nest in a sticky ball moistened with regurgitated honey. (Redrawn from Dade, 1977.)

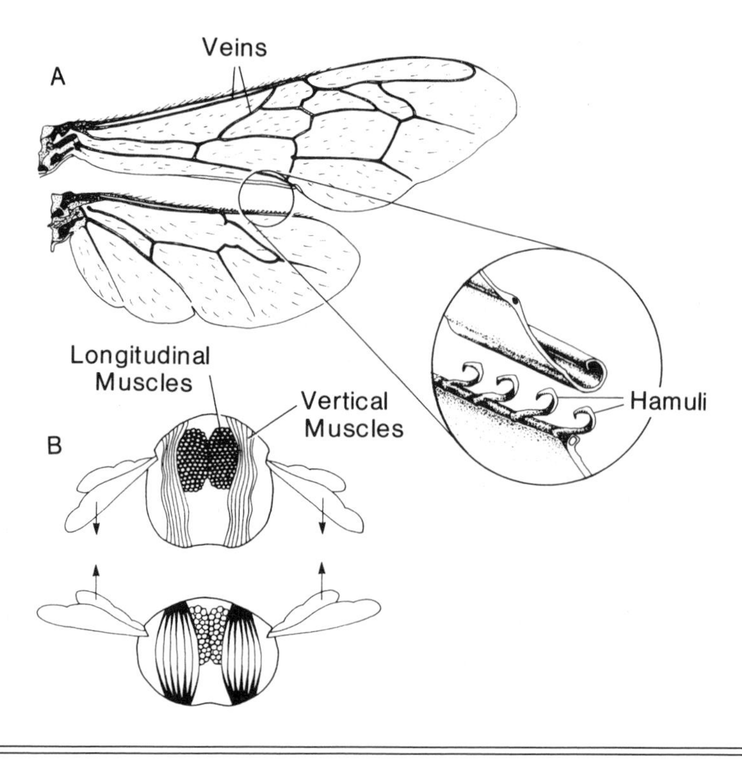

Fig. 3.13

(*A*) The front and hind wings of workers, showing the veins and hooks, or hamuli, which connect the hind wings to folds on the front wings during flight. (*B*) The thoracic musculature, which provides most of the force for flight. The contraction of the longitudinal muscles and the relaxation of the vertical muscles extends the thorax vertically and drives the wings downward. In contrast, the relaxation of the longitudinal muscles and contraction of the vertical muscles bows the thorax outward, driving the wings upward. (Part *b* redrawn from Dade, 1977.)

honey bees can maintain thoracic flight temperature at an extraordinarily high 46°C by passing excess heat to the head by means of passive conduction and accelerated blood flow. This excess heat is then eliminated by regurgitating droplets of watered-down honey, which cools the head much like sweat in vertebrates. This prevents overheating while allowing the thoracic muscles to operate at peak efficiency (Heinrich, 1979b, 1980a,b).

The average flight speed of a worker bee is about 24 km/hr (Park, 1923a; von Frisch, 1967a); workers with full nectar loads fly at about 6.5 m/sec, whereas unloaded workers may fly at 7.5 m/sec (Wenner, 1963). Flight speed and distance depends on the powerful thoracic muscles being amply supplied with energy derived from nectar metabolism; if the blood sugar falls below 1% the bee can no longer fly

(Dade, 1977). For fuel, workers engorge with honey and store it in their stomachs before leaving the nest, gradually consuming it during flight. In studies of swarm flight with Africanized bees, a typical worker carried 30 mg of honey, which contained about 20 mg of sugars (Otis, Winston, and Taylor, 1981). With this average honey load, a worker could fly about 60 km before running out of fuel.

The Abdomen

The abdomen of a worker is made up of seven visible segments, counting the propodeum as the first abdominal segment. Two additional segments can be found associated with the worker sting or the reproductive organs of the queen and drone, but these are highly reduced and appear internally only as small, soft plates. Each abdominal segment is constructed of one large dorsal and one ventral plate, the dorsal overlapping the ventral, and the two plates connected by membranes. These connective membranes are important because they allow the abdomen to expand when the stomach is engorged with nectar or water and also permit pumping of the abdomen, which increases oxygen intake during active periods. The abdomen is generally hairy but otherwise has no external structures of great interest, with the exception of the sting, which is held inside a chamber at the tip of the abdomen. Internally, the abdomen contains most of the organ systems and some glands.

The worker sting is a highly modified ovipositor which has evolved for its defensive functions (Fig. 3.14). Unlike most stinging insects, the honey bee loses its sting after use, resulting in the bee's death shortly afterward. The advantage of losing its sting is that the victim is injected with additional venom. In colonies which may have thousands of workers, the loss of a few during colony defense is thus balanced by the added effectiveness of extra venom.

The sting is made up of two barbed lancets supported by hardened plates and strong muscles and connected to a poison gland and glands containing alarm substances. When the worker stings, the lancets quickly saw their way into the victim, and the forward-facing barbs catch hold in the skin. The worker then pulls away, ripping the sting out, and dies within a few hours or days as a result of the massive abdominal rupture (Haydak, 1951). The sting remains behind and continues to pump venom for 30–60 sec, as the muscles surrounding the poison sac contract. Associated glands give off alarm chemicals that induce other workers to sting.

The venom injected when a worker stings is a mixture of proteins and peptides, the major component being a protein called melittin. Venom contains other compounds such as hyaluronidase, phospho-

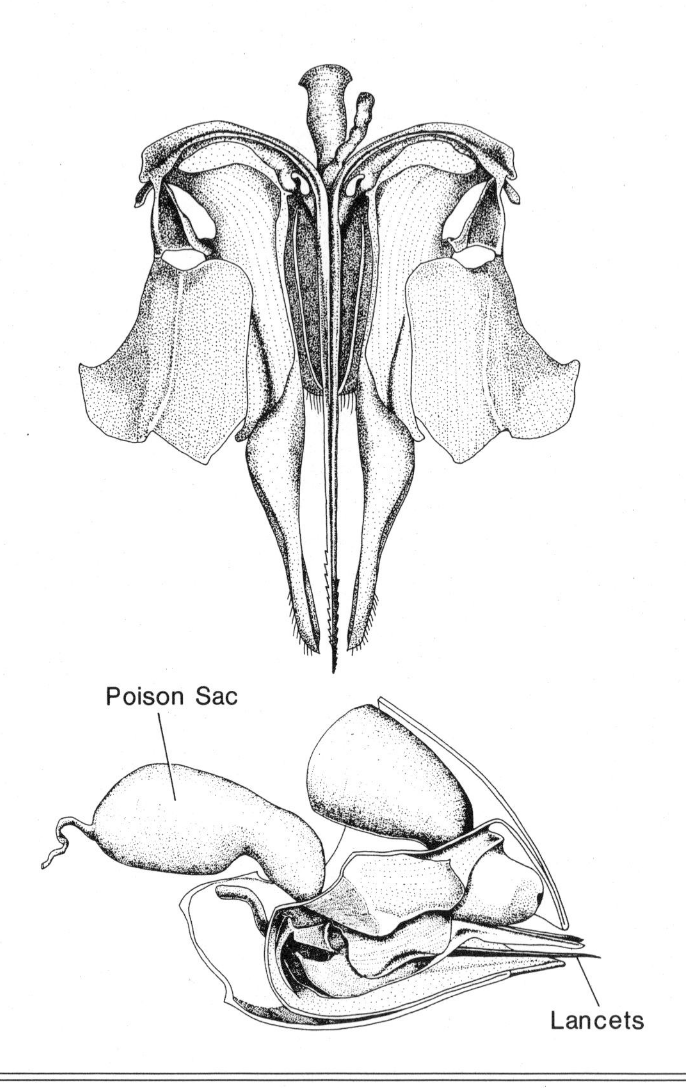

Fig. 3.14

Side and front views of the sting, showing the poison sac, barbed lancets, and associated muscles and hard plates. (Redrawn from Dade, 1977, and Snodgrass, 1956. Copyright © 1956 by Cornell University. Used by permission of Cornell University Press.)

lipase A, acid phosphatase, and histamine (Habermann, 1971; Owen, 1978a,b; Riches, 1982; Schmidt, 1982). The complex nature of the venom may be due to the wide variety of insect and vertebrate pests and predators which might attack a bee colony; different components of the venom seem to be important in repelling different species of attackers. For example, the amount of histamine in a bee sting is not toxic to vertebrates but is a significant part of toxicity against other insects, including other honey bees (Owen, 1978a,b). Against verte-

Table 3.1 Effects of bee venom components on vertebrates

Compound	Percentage in venom (dry weight)	Effects
Melittin	50	Lysis of blood and mast cells Release of histamine and serotonin from mast cells Depression of blood pressure and respiration
Phospholipase A	12	Cell lysis Pain Toxicity; synergistic with melittin
Hyaluronidase	<3	Hydrolyzes connective tissue; called "spreading factor" since it opens up passages for other components Nontoxic
Acid phosphatase	<1	Involved in allergic reaction
Histamine	<1	Itching Pain Amount in venom much lower than toxic levels or amount released by mast cells

Source: Habermann, 1971, and Schmidt, 1982.

brates, each of the major venom components has somewhat different effects, the sum of which we see as an allergic reaction (Habermann, 1971; Schmidt, 1982) (Table 3.1).

In humans reactions to stings take place on three levels: local, systemic, and anaphylactic. In the first kind of reaction the initial localized swelling is followed by more extensive swelling a few hours later, and the affected area may be red, itchy, and tender for 2–3 days. A systemic reaction generally occurs within a few minutes of stinging, and it may involve a whole-body rash, wheezing, nausea, vomiting, abdominal pain, and fainting. In an anaphylactic reaction symptoms can occur within seconds, and they include difficulty in breathing, confusion, vomiting, and falling blood pressure, which can lead to loss of consciousness and death from circulatory and respiratory collapse (Frankland, 1976; Riches, 1982). Generally, one can develop some resistance to bee stings the more one is stung, although the reaction to stings can become suddenly acute for no apparent reason. Those who are extremely sensitive may die from a single sting. Yet, the case has been reported of a man receiving 2243 stings and surviving (Murray, 1964).

Internal Systems

DIGESTIVE AND EXCRETORY SYSTEMS

The digestive system of honey bees is located primarily in the abdomen, connected to the mouth via the long esophagus (Fig. 3.15). The posterior end of the esophagus opens into the crop or honey stomach, an expandable bag which holds honey ingested in the hive and used for energy during flight, and nectar or water collected in the field by workers for transport back to the nest. When full, the crop occupies much of the abdominal cavity, which expands by stretching the connective membranes that hold the abdominal sclerites together. The contents of the crop can be regurgitated when the surrounding muscles contract and the abdominal segments telescope, squeezing the crop contents back out the esophagus through the mouth to the tongue.

A valve at the end of the crop, the proventricular valve, prevents

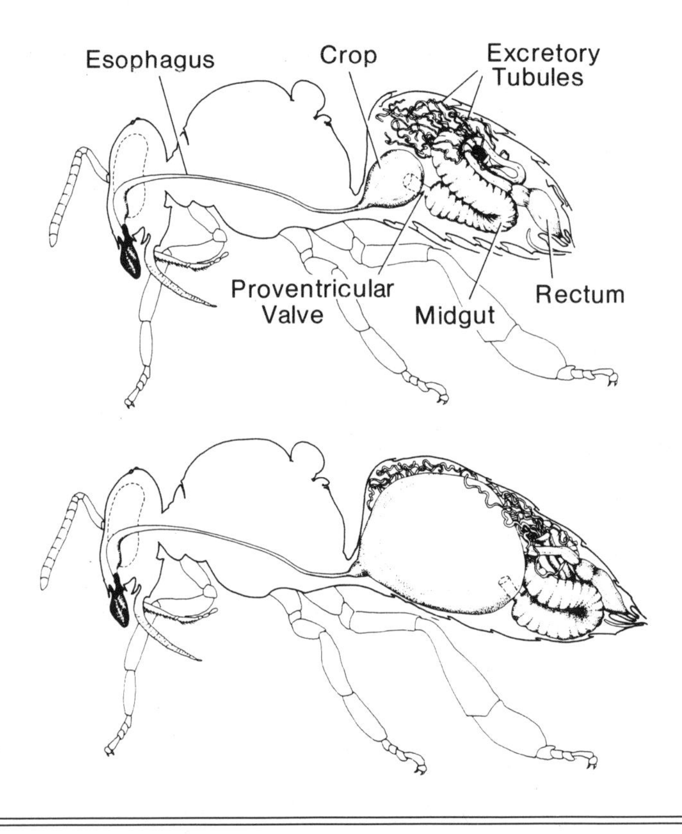

Fig. 3.15

The digestive and excretory system of a worker bee, shown with the crop empty (*top*) and fully engorged (*bottom*). (Redrawn from Michener, 1974, and Dade, 1977.)

most of the liquid crop contents from passing through to the ventriculus, or midgut. Pollen grains, however, are filtered from the crop along with some of the liquid crop contents and passed onto the midgut, where most of the digestion and absorption take place (Bailey, 1952; Dietz, 1969). Solid wastes, consisting primarily of pollen husks, fat globules, and dead midgut cells, are then passed through the intestine to the rectum for excretion. Liquid nitrogenous wastes are absorbed from the blood by the Malpighian tubules and passed to the intestine for excretion. The rectum expands considerably to hold waste material during the winter, since bees usually do not defecate in the hive and must wait for warm flying weather in the spring to eliminate accumulated feces.

For storage of food reserves, bees have cream-colored cells on the dorsal and ventral parts of the abdomen called fat bodies. These cells concentrate and store fat, protein in the form of albumen, and glycogen, which can be quickly converted to glucose when needed.

CIRCULATORY, RESPIRATORY, AND NERVOUS SYSTEMS

In insects the circulatory and respiratory systems are separate, and the blood has only a minor role in gas transport to and from the cells. The bee circulatory system is an open one, with only a dorsal heart and aorta to assist in blood circulation (Fig. 3.16). The blood fills the

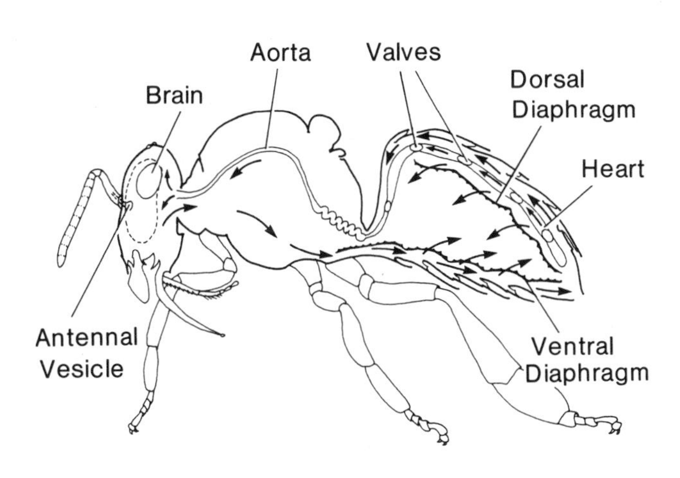

Fig. 3.16

The worker circulatory system. Blood enters the heart through valves, is pumped anteriorly through the aorta, empties into the body cavity near the brain, and is pumped posteriorly by muscles associated with the diaphragms. The antennal vesicles and similar structures at the bases of the legs and wings assist in pumping blood to the extremities. (Redrawn from Dade, 1977.)

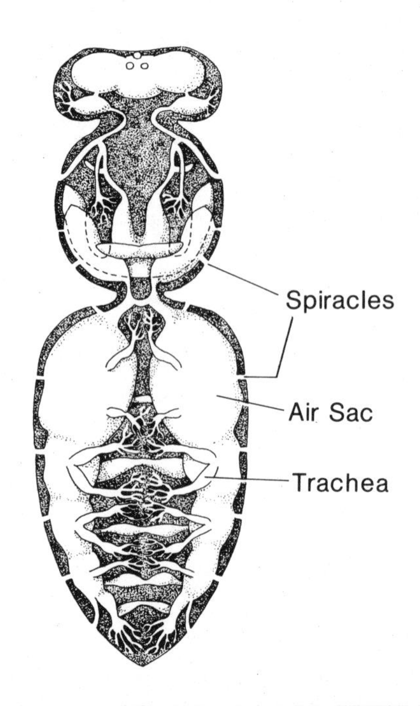

Fig. 3.17

The vascular ventilatory system, made up of spiracles, through which air enters and leaves the worker's body, and the main tracheae and air sacs, which carry air to and from the cells. (Redrawn from Dade, 1977.)

bee's body cavity, so that cells float freely in blood rather than receive blood through vessels. The blood enters the heart through one-way valves called ostia and is pumped forward through the aorta, from which it spills out into the head. Muscles attached to a dorsal and ventral diaphragm are used to pump the blood throughout the body and back to the heart. The main functions of the circulatory system are transporting food from the midgut to the body cells, removing waste material from cells and returning it to excretory organs, lubricating body movements, and providing defense against pathogens by means of blood cells that attack invading organisms.

Bees have no lungs for breathing; rather, they use a system of tubes which carry oxygen to and carbon dioxide away from cells (Fig. 3.17). This system of breathing tubes, or tracheae, is connected to the outside world by a series of holes in the cuticle called spiracles. When the bee is inactive gas exchange can operate simply by diffusion, but during periods of increased activity bees pump their abdomens to increase gas exchange, using expanded sacs of the trachea as bellows (Bailey, 1954).

The gross structure of the nervous system is a fairly simple one, consisting of a brain and seven ganglia, or nerve centers, at various junctions throughout the body (Fig. 3.18). Much of the nervous control exercised by bees is performed not by the brain but by these centers, which provide local control over some of the bee's musculature. For example, a beheaded bee can still beat its wings, move its legs, and sting, although coordinated activity of these functions has been lost.

GLANDULAR SYSTEMS

The glands of worker honey bees are used for four basic functions: wax production, communication, defense, and food processing. A brief summary of the gland structures and functions as far as they are known is given below. In spite of research dating back to the early 1800's, however, many of the functions of and chemicals produced by these glands are still not well understood.

Wax production. Wax used for comb construction, or beeswax, is produced by modified epidermal cells located ventrally on the fourth to seventh abdominal segments (Fig. 3.19). These paired glands are concealed beneath overlapping plates, called wax mirrors, on each segment. The wax-secreting cells shrink when wax is not being pro-

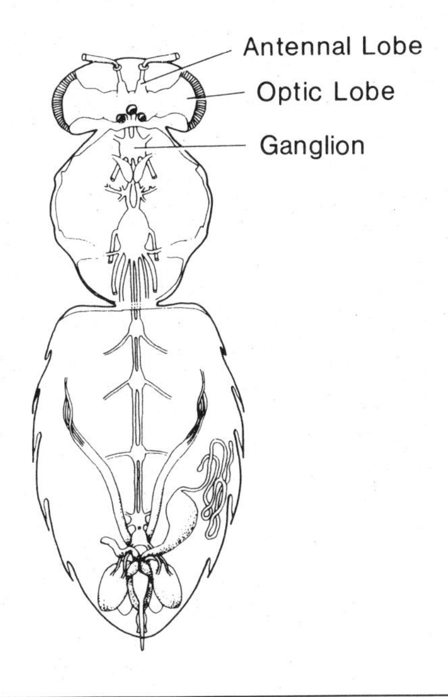

Fig. 3.18

The nervous system of a worker bee, showing the brain and the ganglia, or nerve centers. (Redrawn from Dade, 1977.)

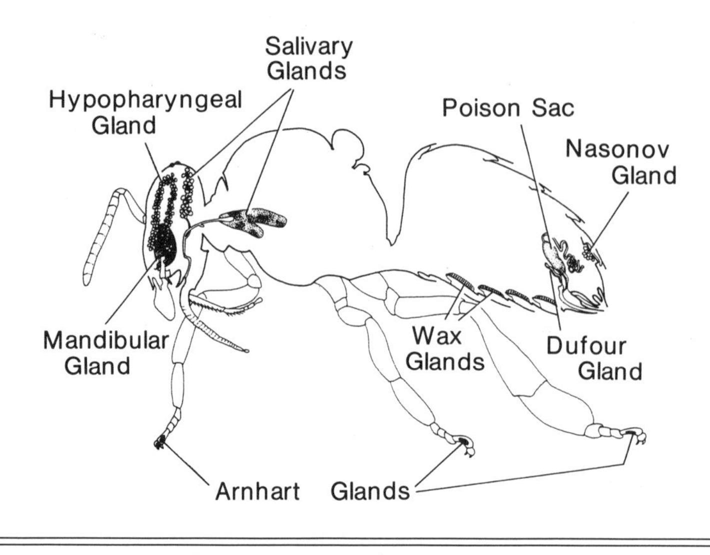

The worker glandular system. (Redrawn from Michener, 1974.)

duced but enlarge when the worker is producing wax (Dreyling, 1903; Rösch, 1927; Boehm, 1965; Cruz-Landim, 1963). The liquid wax secreted from the glands collects on the wax mirrors and hardens into visible scales which are removed by brushing the spines on the hind basitarsi past the mirrors (Casteel, 1912; Rösch, 1927). The leg holding the scale moves forward to the mandibles, which then manipulate the wax with the help of the forelegs for comb construction.

The wax used for all of this construction is a substance unique to social bees. It is produced by metabolizing honey in fat cells associated with the wax glands and converting it to beeswax; workers cannot produce beeswax unless there are adequate honey stores in the colony. It takes 8.4 kg of honey to produce the 991,000 wax scales which make up 1 kg or 2.2 lb of wax (Whitcomb, 1946). Also, workers need to eat pollen during the first 5–6 days of their life in order to secrete wax later on, evidently because the protein in pollen is needed at that time for adequate fat cell development (Goetze and Bessling, 1959; Freudenstein, 1960).

The chemical composition of beeswax has been reviewed by Callow (1963) and Tulloch (1980) and consists of a blend of over 300 individual components, primarily hydrocarbons (14%), monoesters (35%), diesters (14%), hydroxy polyesters (8%), and free acids (12%). Beeswax is invariably white in color, although it generally contains yellow hues caused by fat-soluble carotenoid pigments that originate from pollen (Vansell and Bisson, 1935; Tischer, 1940; Freudenstein, 1962).

Communication. The known glands in workers which dispense odors used in communication include the Nasonov gland, the mandibular glands, the setose sting glands, and possibly the tarsal Arnhart glands. The Nasonov, or scent, gland is found beneath the tergite of the last abdominal segment, which secretes its products into a duct, the scent canal (McIdoo, 1914; Jacobs, 1924; Renner, 1960; Belik, 1979). The chemicals produced by this gland are released when the worker uncovers the scent canal, raises the abdomen, and fans the wings, dispersing the Nasonov chemicals. The uncovered scent canal is visible as a light brown band across the abdomen; when workers are fanning at the entrance for ventilation, the abdomen is not raised, and the scent gland remains covered and cannot be seen.

The Nasonov scent is actually a mixture of seven chemicals, including geraniol, nerolic acid, geranic acid, (*E*)-citral, (*Z*)-citral, (*E,E*)-farnesol, and nerol (Pickett et al., 1980). Of these, geraniol, nerolic acid, and (*E,E*)-farnesol are present in the highest proportions; however, a mixture of all components is as or more attractive than the natural blend (Free et al., 1984). The Nasonov scent is used for orientation, particularly at the nest entrance (Butler and Calam, 1969), in swarm clustering (Morse and Boch, 1971; Avitabile, Morse, and Boch, 1975), at water collection sites (Free and Williams, 1970), and possibly at flowers (Free and Williams, 1972; Free, Ferguson, and Pickett, 1983; Waller, 1970).

The paired mandibular glands are found on each side of the bee's head, each attached to its mandible by a duct. These are large glands, extending to the base of the antennae in the workers, and release of their active products is controlled by a valve which allows the bee to regulate the discharge of secretions from a storage sac (Nedel, 1960). In younger workers that are producing larval food, these glands are involved in brood food production, particularly 10-hydroxy-2-decenoic acid, which is the main lipid component of larval food (Butenandt and Rembold, 1957), and octanoic and other volatile acids (Boch, Shearer, and Shuel, 1979). But this secretion changes in older workers (Costa-Leonardo, 1980) so that the glands produce 2-heptanone, an alarm substance (Shearer and Boch, 1965). Although 2-heptanone will excite workers, it is not nearly as active as the sting-produced compounds (Boch, Shearer, and Petrasovits, 1970; Gary, 1974).

The sting-produced complex of pheromones can be extremely effective in eliciting defensive behavior. The exact location of the glands which produce these compounds is not known, but they appear to be located in a setose membrane at the base of the sting (Maschwitz, 1964a,b; Butler, 1967) and are released when the bee everts the sting and fans or when the sting is torn from the worker (Maschwitz,

1964b). For many years only one compound, isoamyl acetate, was identified as a sting alarm substance (Boch, Shearer, and Stone, 1962), which functions to excite workers and attract them to a potential victim (Ghent and Gary, 1962). More recently, a considerable number of additional compounds from the sting have been identified which also appear to elicit similar alarm behavior, including 2-nonanol, *n*-butyl acetate, *n*-hexyl acetate, benzyl acetate, isopentyl alcohol, *n*-octyl acetate, and (Z)-11-eicosen-1-ol (Blum et al., 1978; Collins and Blum, 1982, 1983; Pickett, Williams, and Martin, 1982). However, such a large number of alarm compounds is unusual in a social insect, and some of these substances may have other functions.

The last tarsal segments of all six worker legs contain the Arnhart glands, which empty their contents through the tarsal pads (Arnhart, 1923; Chauvin, 1962). Their function is not known, although it has been suggested that these glands produce the "footprint" substance which workers may deposit at the nest entrance or at flowers to guide incoming foragers (Butler, Fletcher, and Watler, 1969; Ferguson and Free, 1979).

Defense. The major gland with a defensive function is the poison gland, the large sac associated with the sting which holds the venom. This gland has been called the acid gland because of the mistaken impression that it contained formic acid (Snodgrass, 1956; Michener, 1974). It consists of cells that secrete the venom into the poison sac, which is surrounded by muscles that pump the venom through the sting (Cruz-Landim and Kitajima, 1966; Bridges, 1977). Another small gland which discharges its content into the sting chamber is the small alkaline, or Dufour, gland, of unknown function in honey bees. Various functions have been suggested for it, including poison secretion, sting lubrication, secretion of a waxy covering for the egg, and attachment of eggs to the bottom of cells (Trojan, 1930; Kerr and Lello, 1962).

Food processing. There are two types of food processing glands, those which partly digest food and those involved in brood food production. The digestive type are the labial, or salivary glands, including a gland in the posterior part of the head as well as the thoracic labial gland. Both connect through a common salivary duct to the mouth, and they function to dissolve sugars, clean the queen, and possibly soften material which needs to be chewed. The secretion of the thoracic glands is a watery saliva which dissolves sugars, whereas the head glands produce an oily secretion of unknown function (Simpson, 1960; Arnold and Delage-Darchen, 1978). The size of both salivary glands does not vary with worker age, and pollen consumption by adult bees is apparently not important for gland functioning (Kratky, 1931).

The hypopharyngeal glands produce some of the proteins, lipids, and vitamins in the food fed to larvae by adult workers (Patel, Haydak, and Gochnauer, 1960), and they also secrete the enzyme invertase, which is important in processing floral nectars into honey (Simpson, 1960; Simpson, Riedel, and Wilding, 1968), and an enzyme that oxidizes glucose to an acid (Cruz-Landim and Hadek, 1969). These paired glands are located behind the face, with a duct opening at the base of the tongue (Cruz-Landim and Hadek, 1969; Ortiz-Picon and Diaz-Flores, 1972). They are relatively large in younger workers producing brood food, and protein consumption is necessary for the glands to develop fully. In older workers the glands resorb when nursing duties are completed, although they may expand again if nursing is required of older bees (Rösch, 1930; Moskovljevic, 1940; Halberstadt, 1980). After resorption, they primarily produce invertase (Simpson, Riedel, and Wilding, 1968). The mandibular glands also produce some of the brood food, notably 10-hydroxy-2-decenoic acid.

Queen glands. Two other glands found in honey bees are most developed in queens, though neither has had its functions adequately described. The Koshevnikov gland is associated with the queen's sting and may also occur in a smaller form in workers; several studies have suggested that it produces attractant scents (Altenkirch, 1962; Butler and Simpson, 1965; Hemstedt, 1969). Epidermal glands are found on all parts of the bee's body, particularly the abdomen. Their functions are not yet known, but they may also produce attractant substances (Heselhaus, 1922; Jacobs, 1924; Snodgrass, 1956; Renner and Vierling, 1977; Vierling and Renner, 1977). In addition, the Dufour gland is well developed in queens, possibly because of their egg-laying functions.

The mandibular glands are also very large in queens, and they produce at least 2 substances important in regulation of hive activities, the so-called queen substance 9-keto-(*E*)-2-decenoic acid (Barbier and Lederer, 1960; Callow and Johnston, 1960) and 9-hydroxy-(*E*)-2-decenoic acid (Butler, Callow, and Chapman, 1964). The activity of these compounds may include inhibition of queen rearing, swarming, and worker egg laying, attraction of drones for mating, stimulation of the release of the Nasonov pheromone, colony recognition, and worker orientation. There are numerous other compounds produced in the queen's mandibular glands, the functions of which are not known (Callow, Chapman, and Paton, 1964).

Specialized Functions in Drones and Queens

One of the principal advantages to having individuals of different physical castes in a social insect colony is that each caste can be specialized for particular functions. In honey bee colonies drones and

Table 3.2 Some structural differences between worker, queen, and drone castes in honey bees

Characteristic	Worker	Queen	Drone
Sensory			
No. facets of compound eyes	4000–6900	3000–4000	7000–8600
Optic lobe of brain	Medium	Small	Large
No. antennal plate organs	3000	1600	30,000
Relative ratio of antennal surface	2	1	3
Glandular			
Hypopharyngeal	Present	Vestigial	Absent
Mandibular	Large	Very large	Small
Head salivary (labial)	Large	Large	Vestigial
Thoracic salivary (labial)	Large	Large	Small
Wax	Present	Absent	Absent
Nasonov	Present	Absent	Absent
Alkaline (Dufour)	Reduced	Large	Absent
Koshevnikov	Reduced or absent	Present	Absent
Reproductive and sting			
Ovaries or testes	Reduced ovaries	Enlarged ovaries	Testes
No. ovarioles	2–12	150–180	None
Spermatheca	Rudimentary	Large	None
Sting barbs	Strong	Minute	No sting
Sting plates	Loosely attached	Strongly attached	None
Mouthpart			
Mandibles	Slender	Robust	Small
Mandibular groove	Present	Absent	Absent
Proboscis	Long	Short	Short
Leg and Wing			
Pollen press and combs	Present	Absent	Absent
Pollen basket	Present	Absent	Absent
Wing sensilla	Medium	Fewest	Most

Source: Data from Ribbands, 1953, Snodgrass, 1956, Michener, 1974, Dade, 1977, and other references cited therein.

queens do not perform such worker jobs as brood rearing, comb building, and foraging, and thus do not have many of the worker structures used for these tasks. Conversely, both drones and queens have some unique specializations, particularly those associated with reproduction (Table 3.2).

DRONES

Drones are designed for their only significant function, mating. Since a drone performs no work for the colony and is fed by workers, the drone's work-related structures are reduced or absent. For example, drones have a relatively short proboscis; vestigial or very small head salivary glands; no wax, hypopharyngeal, or Nasonov glands; small mandibles; a slender crop; no pollen-collecting structures on the legs; and no sting. In contrast, orientation, flight, and mating-related structures are highly developed. Their compound eyes are much larger than workers', taking up much of the head, and with up to 8600 facets as opposed to 6900 for workers (Dade, 1977). The optic lobes of the brain are also much larger in drones, and their antennae have 30,000 olfactory plate organs, about ten times the number found in workers. These antennal and eye characteristics are presumably important in visual and olfactory orientation for finding and mating virgin queens. Drones also have larger flight muscles and broader wings than workers, both adaptations for mating flights. The mandibular glands are relatively small but may produce pheromones used for orientation (Lensky et al., 1985).

The drone reproductive system is designed to evert the genital apparatus into the queen during mating. The drone penis consists of an internal endophallus, which is everted upon ejaculation, and a pair of copulatory claspers for gripping the queen (Fig. 3.20). Drones mate only once and then die, since much of the endophallus breaks off from the drone and is left in the queen during copulation. The sperm transferred during mating are produced in the testes, which reach their largest size during the pupal stage; sperm production is largely completed by the time the adult drone is mature, about 12–13 days after emergence (Zander, 1916; Bishop 1920a,b). The sperm are stored in the seminal vesicles until mating, when they are ejaculated in a packet with mucus from the large mucus glands.

QUEENS

Many of the task-related structures found in workers are also reduced or absent in the queen. The queen's proboscis is shorter, she lacks pollen-collecting structures, and her wax, hypopharyngeal, and Nasonov glands are absent or vestigial. Queens do have stings, which are used in battles against rival queens, but the barbs are small and ineffective, and the associated plates firmly attached, so that the sting can be retracted after use. Thus, queens generally do not die after stinging. Their poison sacs contain two to three times the volume of worker venom (Owen, 1978b), possibly because the queen may need to sting many rivals in a short time. The queen also has some well

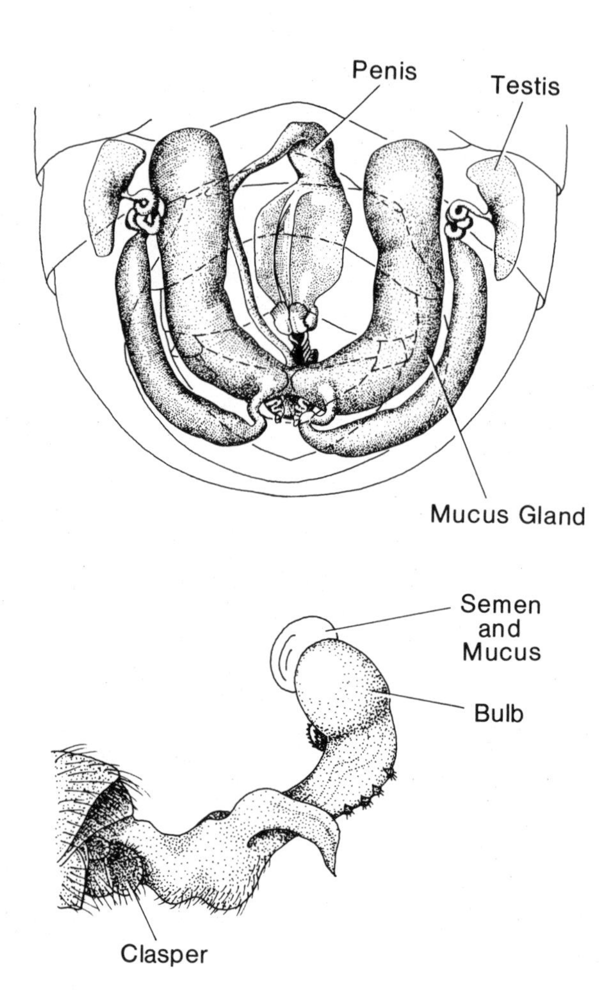

Fig. 3.20

The major structures of a drone's reproductive system, shown held inside the abdomen prior to mating and following extension for mating. (Redrawn from Dade, 1977, and Snodgrass, 1956. Copyright © 1956 by Cornell University. Used by permission of Cornell University Press.)

developed glands for pheromone production, particularly the large mandibular glands and possibly the Koshevnikov glands.

The queen's ovaries are enormous compared to the worker's, obviously because of her egg-laying function (Fig. 3.21). Each of her two ovaries consists of 150–180 egg-producing ovarioles, whereas a worker ovary has only 2–12. These ovarioles can produce an unlimited number of eggs, often as many as one million or more during the queen's lifetime. The eggs travel down the oviducts past the spherical spermatheca to the vagina, from where they are laid in cells. The spermatheca holds sperm from the drones which have mated with the queen early in her life, and nutrients supplied by the spermathecal gland ensure that the sperm will survive for many years.

When an egg passes down the oviduct, it is probably pressed by a muscular flap, the valve fold, against the duct opening to the spermatheca (Ruttner, 1956a). A small S-shaped pump and valve leading from the spermatheca to this duct opening allows the queen to draw a minute amount of sperm and seminal fluid into the duct, thus re-

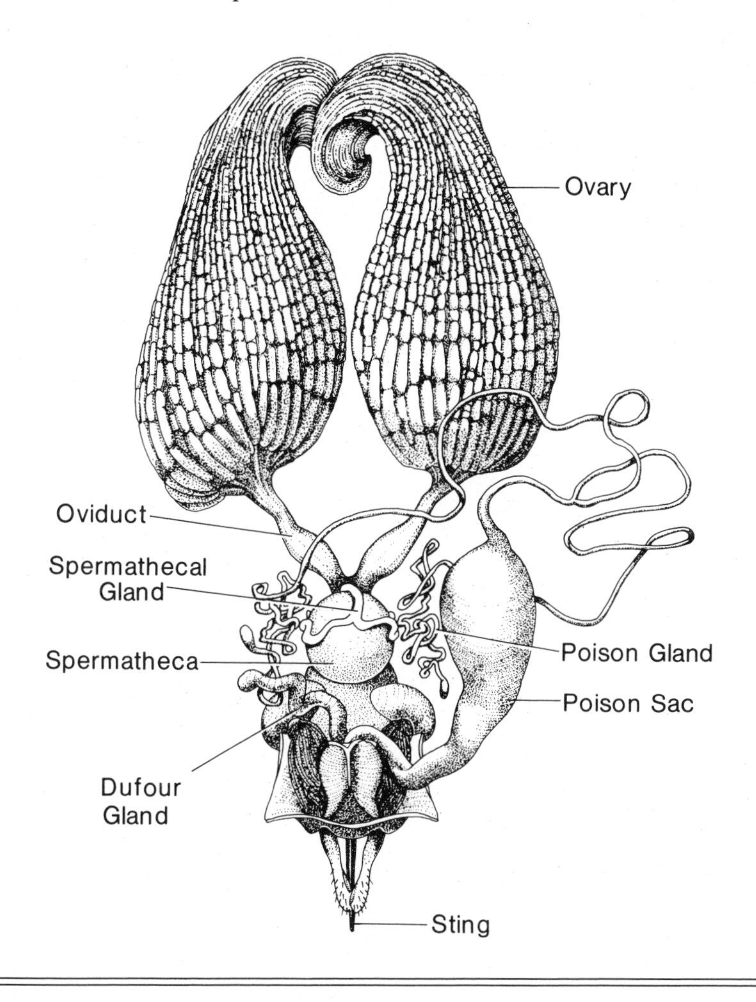

Fig. 3.21

The reproductive system and sting of a mated queen. (Redrawn from Dade, 1977, and Snodgrass, 1956. Copyright © 1956 by Cornell University. Used by permission of Cornell University Press.)

leasing only a few sperm at a time (Bresslau, 1905; Adam, 1912; Harbo, 1979) (Fig. 3.22). This is important to the queen, since she will be superseded and killed by the colony when she runs out of sperm. The spermatheca can hold up to seven million sperm (Dade, 1977), and it generally takes 2–4 yr after mating before all of the sperm are used. Thus, it is the amount of sperm held in the spermatheca, rather than egg production, which determines the natural life span of queens. Workers have a rudimentary, nonfunctional spermatheca and also lack the various genital structures with which the queen can mate and accept sperm from drones (Fig. 3.23). Thus, workers cannot mate or store sperm, and any worker-laid eggs cannot be fertilized.

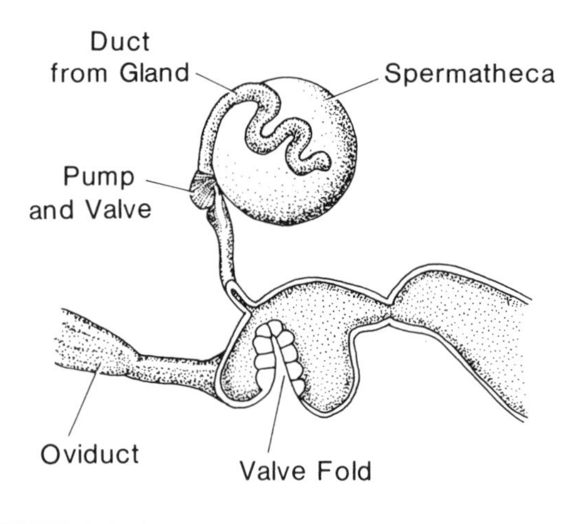

Fig. 3.22

View of the spermatheca and oviduct, showing the pump and valve of the spermathecal duct and the valve fold of the oviduct, which regulate and coordinate sperm release and egg passage. (Redrawn from Dade, 1977.)

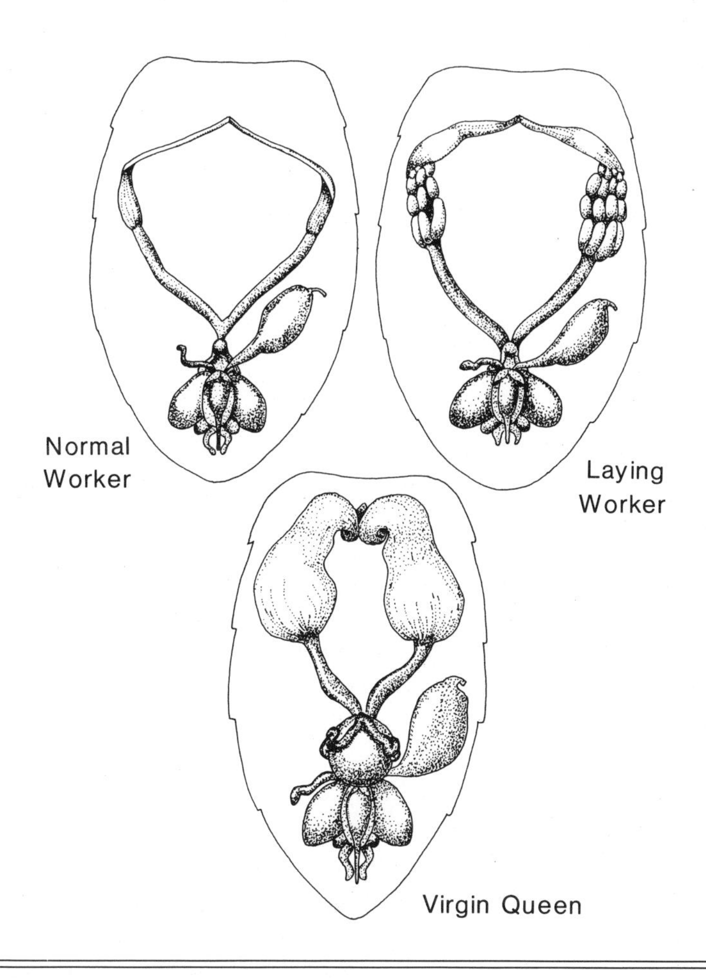

Fig. 3.23

The ovaries of a normal worker, a laying worker, and a virgin queen. (Redrawn from Dade, 1977.)

4

Development and Nutrition

The process of growth and metamorphosis in bees involves some of the most complex interactions in a social insect colony. At the simplest level bee development can be examined from the perspective of a single bee, which must undergo a number of stages before finally emerging as an adult. But the metamorphosis from egg to adult is not unique to honey bees, and most insects go through a similar process. What separates social insects like honey bees from more solitary insects are the interactions that occur between the brood and adults, and it is these relationships that express many of the unique aspects of honey bee societies.

The Life Cycle

Development of all three castes involves a transition through four major stages: egg, larva, pupa, and adult (Fig. 4.1). The queen lays eggs in worker or drone cells; fertilized eggs can develop into either workers or queens, whereas unfertilized eggs usually develop into drones. The larval stage is the feeding time, when the bee gains an enormous amount of weight and grows tremendously in size. These two changes occur while cells are uncapped; the larvae spin their cocoons and change into pupae after adult workers have capped their cells. The pupal stage is one of metamorphosis, when the brood change into adults; when this transformation is complete, the teneral adult chews its way out of its cells and finishes developing during the next few days. The whole process from egg to adult can take as little as 16 days for queens, or as long as 24 days for drones. The development time as well as the quality of the emerged adult are particularly dependent on temperature, nutrition, and the race of bee. The development of bees in their cells has been reviewed by Jay (1963a); unless otherwise indicated, the data cited below have been drawn from that publication and references cited therein.

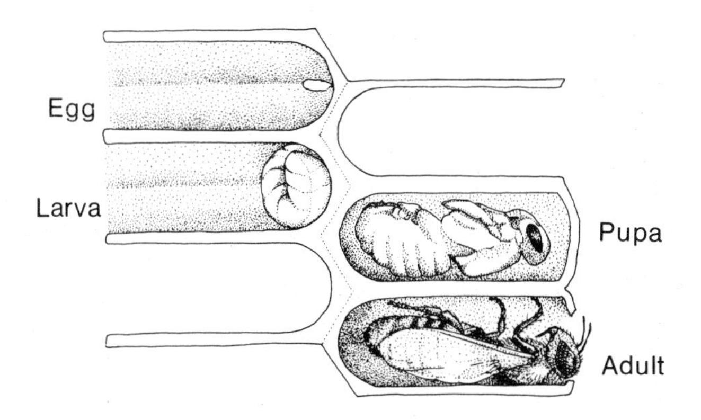

Fig. 4.1

The four major stages in honey bee development. (Redrawn from Sammataro and Avitabile, 1978, *The Beekeeper's Handbook.* Copyright © 1978 by Diana Sammataro and Alphonse Avitabile. Used by permission of Charles Scribner's Sons.)

EGGS

The eggs laid by honey bee queens are small for bee eggs and nondescript, considering the elaborate adults into which they will develop. They are a pearly white color, cylindrical and elongate-oval in shape, and slightly curved, with the end that is to develop into the head somewhat thicker than the abdominal end. Internally, the egg consists of the original egg cell and a large quantity of yolk, which is deposited by the queen prior to egg laying.

When the queen first lays an egg, she glues it to the floor of the cell at one end so that it appears to be standing in the bottom of the cell. Usually only one egg is laid per cell. During the approximately 3-day period before hatching, the egg gradually sags until it finally rests on the cell floor. The hatching of the egg into the first larval stage is almost indiscernible, and the larva slowly becomes exposed as the embryo moves and the egg membrane dissolves. All other insects hatch from their eggs by rupturing the membranes; gradual dissolution of the membrane during eclosion appears to be unique to honey bees (DuPraw, 1961, 1967).

There is considerable variability in egg size and development time, both of which have genetic and environmental components. For worker and queen egg size, weights in the range of 0.12–0.22 mg have been reported (Taber and Roberts, 1963; Roberts and Taber, 1965; DuPraw, 1967), as have egg lengths in the range of 1.3–1.8 mm. Some of this variation can be explained by heritable differences between queens, and hybrid crosses between high and low weight lines result in eggs of intermediate weight (Taber and Roberts, 1963). But even eggs from a single queen can vary tremendously. DuPraw (1961) found that one egg laid by a queen was almost half the size of an egg she had laid 90 min earlier, although both developed normally and hatched almost simultaneously. Also, eggs lose about 30% of their weight during incubation, mostly due to water loss (DuPraw, 1967).

Development time from egg laying until hatching also varies widely, with values from 48 to 144 hr being reported. Seventy-two hr, however, is generally considered the average incubation time. Tropical-evolved bees have shorter development periods for all stages, including eggs; egg development time for African-evolved honey bees is 70–71 hr on average (Tribe and Fletcher, 1977; Harbo et al., 1981), whereas eggs from European-evolved bees hatch in an average of 72–76 hr (DuPraw, 1961; Harbo et al., 1981).

LARVAE

Bee larvae are essentially feeding machines designed for rapid growth, stripped of all nonessential external parts and equipped with a huge digestive system. The larva is a whitish wormlike grub with no legs, eyes, antennae, wings, or sting, possessing simple mouthparts which need only lap up the copious quantities of food placed in the larval-containing cells by adult workers. Most of the body cavity is taken up by the midgut and hindgut, with the enzyme-secreting salivary glands and the excretory tubules being the other principal internal structures (Fig. 4.2).

Developing bees undergo six molts during which the outer skeleton is shed; five of these take place during the larval stage, and the last occurs when the bee emerges as an adult. The first four larval molts occur approximately once a day for workers and queens and allow the larva to grow rapidly by shedding the exoskeleton when it has become too small. During this time the cells are uncapped, and nurse bees feed the larvae large quantities of brood food by placing it in the cells close to or even on top of the larvae. The larvae are able to rotate within the cells to get to food not placed directly next to their mouths (Lindauer, 1952). At this point the larvae are sealed in their cells with wax cappings constructed by adult workers, sometimes with a little food, which may or may not be eaten.

The last few days of larval life are spent constructing a cocoon within the cell. To spin the cocoon, the larvae uncurl and stretch out fully in the cells with their heads toward the capped end (Jay, 1963b) and begin weaving the cocoon with their spinnerets. The main substance used in the cocoon is silk secreted from what will become the thoracic salivary glands in the adults. The larvae also defecate early in cocoon construction; the excretory tubules and midgut are closed

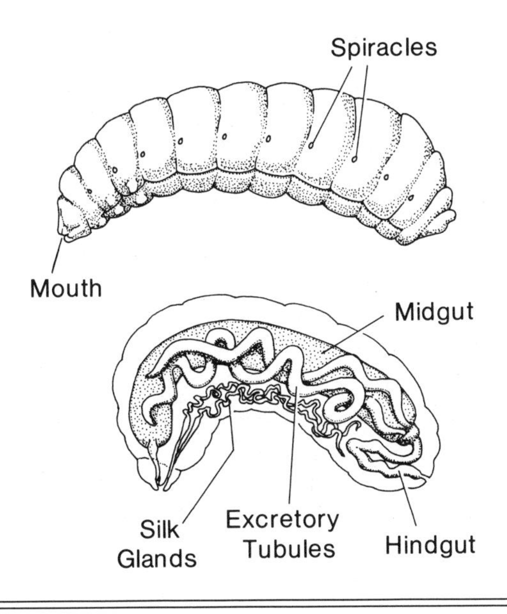

Fig. 4.2

External and internal anatomy of a worker larva. (Redrawn from Dade, 1977.)

off until feeding is completed, when their contents are discharged into the base of the cell. The dark brown feces as well as a lighter-colored substance from the excretory tubules make up most of the other materials used in cocoon construction (Jay, 1964a). This last larval stage is referred to as the prepupal stage, since the final larval molt includes a metamorphosis into the pupae. The prepupal larvae begin to take on the appearance of adult bees just before the cuticle is actually shed, and following the molt the forms of the adult bees are obvious in the pupae (Jay, 1962a). The wings, however, appear as small pads attached to the thorax.

The duration of the larval stages varies between castes and races of bees, with queens showing the shortest larval development times, followed by workers and drones. The development time for larvae is usually considered the duration of the unsealed larval period, since this is easiest to observe. For workers of European races, this is between 5 and 6 days, with an average time of 5.5 days, but with minimum and maximum times of 4 and 11 days, respectively. For tropical African bees, the unsealed worker period is shorter, averaging 4.2 days (Tribe and Fletcher, 1977). For European queens, the unsealed larval stages take 3–5 days, with an average time of 4.6 days, and for European drones the duration of the unsealed larval period is 4–7 days, averaging 6.3 days. After capping, the spinning and prepupal periods last 3–5 days in workers, 3–4 in queens, and 4–6 in drones.

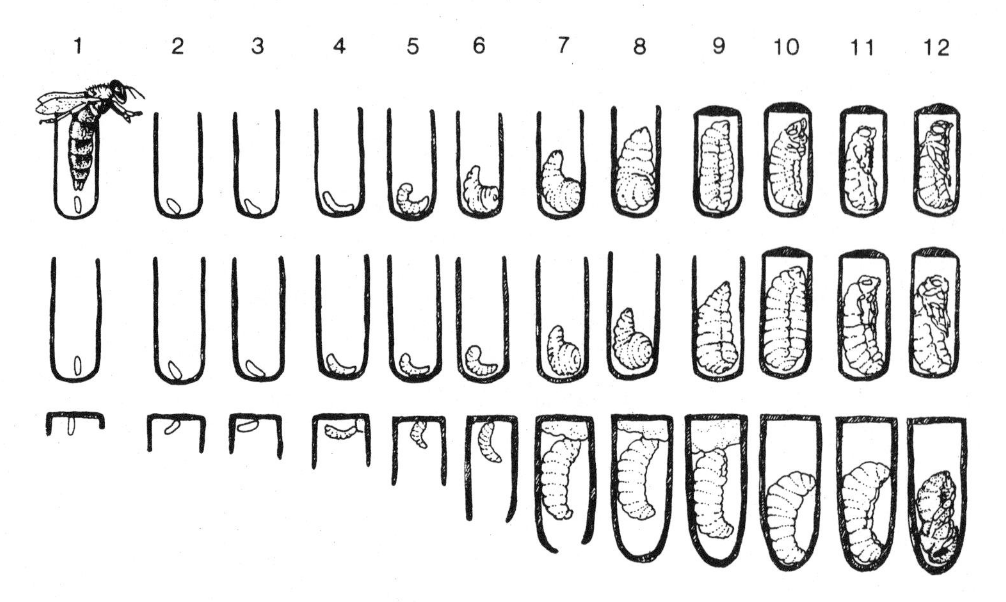

Fig. 4.3

Average developmental times and stages for workers, drones, and queens.

All three castes gain an enormous amount of weight during the larval stage, about 900, 1700, and 2300 times the egg weights for workers, queens, and drones, respectively. Worker weights at capping are approximately 140 mg; queens and drones weigh about 250 and 346 mg, respectively. They also grow in length; on average workers grow from 2.7 to 17.0 mm as larvae, and queens grow from 4.2 to 26.5 mm.

PUPAE

The pupal stage is the last period before the final molt to the adult, and the head, eyes, antennae, mouthparts, thorax, legs, and abdomen all show adult characteristics; only the wings are still small and undeveloped. As the pupa develops, the cuticle gradually becomes darker, and these well-defined color changes can be used to determine pupal age (Jay, 1962b). The pupae do not grow or change shape externally, but internally the muscles and organ systems undergo massive changes into their adult forms. This stage lasts about 8–9 days for workers and drones and 4–5 days for queens, and is followed by the final molt to the adult stage.

Following this final exoskeletal shedding, the teneral adult remains inside the cell for several hours as the new cuticle begins to harden. To emerge, the tenerals begin by using their mandibles to perforate

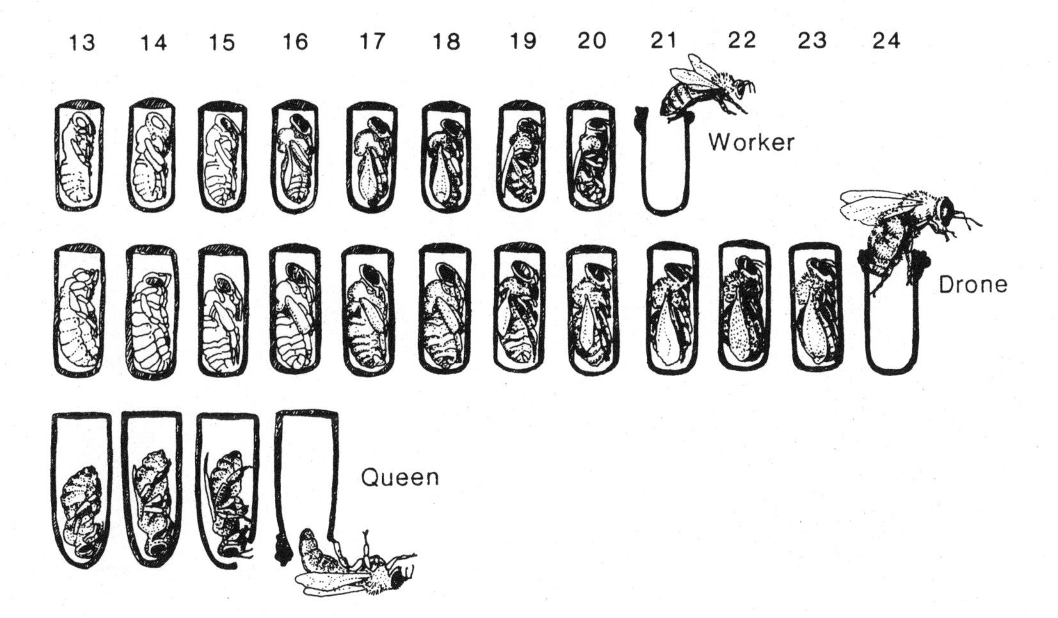

the cell capping with small holes as they rotate within the cell; the antennae frequently protrude through these holes. The wax cappings are manipulated with the mandibles and fastened to the cell wall, where adult workers pick them up and reuse them to cap other cells. After considerable straining and some failed attempts, the emerging bee finally enlarges the cell sufficiently to exit. Once out on the comb, the teneral adult unfolds its wings and antennae, allows the body hairs to dry, and begins its activities.

The total development times from egg laying until adult emergence for European bees generally are considered to be 16, 21, and 24 days for queens, workers, and drones, respectively (Fig. 4.3). However, there is considerable variation in these figures, with ranges for development of 14–17, 16–24, and 20–28 days recorded for the three castes. Much of this variability can be attributed to environmental factors, particularly temperature and nutrition. Temperatures lower than the normal brood nest temperature of 35°C at any stage can delay emergence for up to 5 days, and underfeeding of larvae also will delay development time. Brood developing at the periphery of colonies takes longer to develop than centrally located brood, probably because of problems in maintaining constant temperature and humidity at these locations (Fukuda and Sakagami, 1968).

There is also a genetic basis to development time differences; African-evolved bees have shorter worker and queen development times than temperate-evolved races, averaging 18.5 and 15 days, respectively (Kerr et al., 1972; Tribe and Fletcher, 1977; Fletcher, 1978a; Winston, 1979b). Drones develop in 24 days, the same as for European races (Smith, 1960). The decreased female development time for tropical bees is one of the factors contributing to the more rapid colony growth characteristic of those races (Fletcher, 1978a; Winston, Taylor, and Otis, 1983), which in turn contributes to a higher fecundity for African-evolved bees. Interestingly, hybrid workers between European and African races show intermediate development times, about 20 days from egg to adult, indicating the importance of a heritable factor in determining the duration of the brood stages (Garofalo, 1977).

It is remarkable that under normal colony conditions brood mortality is low, and colonies are able to rear most of the eggs laid by queens to the adult stage (Fig. 4.4). In one study of worker brood survival in colonies of Italian bees (*A. m. ligustica*) in the summer, 94% of the eggs laid survived to the larval stage, 86% to the sealed brood stage, and 85% to the adult stage (Fukuda and Sakagami, 1968). The low mortality rate characteristic of the sealed brood period is due to the brood not requiring feeding and being less sensitive to environmental fluctuations than in the uncapped egg and larval stages. Even higher worker survival rates have been noted, with over 90% and up to 97% of the eggs laid by queens of both European and Africanized races surviving to adulthood (Winston, Dropkin, and Taylor, 1981). Brood survival rate for drones (*A. m. ligustica*) is somewhat lower, averaging 56% from egg to adult, with 82% surviving until the unsealed brood stage, and 60% surviving until the cells are capped. Drone brood at the periphery of the nest has the poorest survival rate, with sealed brood showing the lowest mortality of the brood stages (Fukuda and Ohtani, 1977). Queen brood survival rate is similar to drones, about 53% from egg to adult (91% eggs, 75% larvae; Lee, 1985).

The effect of environmental conditions within the colony on brood survival is confirmed by observations of the brood under stressed conditions, when mortality is much higher. For example, newly hived swarms of Africanized bees showed 32% brood mortality, and mortality following swarming was 42 and 44% for European and Africanized colonies, respectively (Winston, Dropkin, and Taylor, 1981). Colonies which have lost their queens also exhibit high brood mortality, about 44–50% (Winston 1979b; Punnett and Winston, 1983). Under such stressed conditions or even in healthy colonies at the periphery of the brood nest, there are not enough workers, particularly nurse bees, to tend brood adequately and maintain the proper temperature and humidity for optimal development. Even

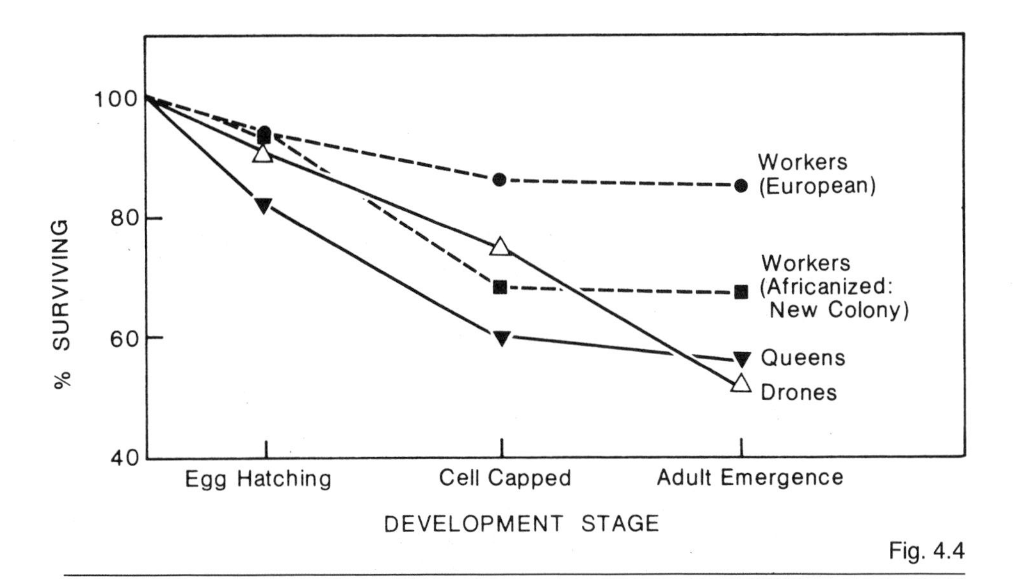

Fig. 4.4

Average proportions of brood surviving until egg hatching, cell capping, and adult emergence for workers, drones, and queens of European races, and for the first Africanized workers to emerge after colony founding by swarms.

brood that die are not wasted, however, since most brood not reared are eaten by workers (Myser, 1952; Fukuda and Sakagami, 1968; Woyke, 1977). Under conditions of pollen dearth, brood cannibalism may be an important source of protein for adult survival until conditions improve (Weiss, 1984).

Other factors in addition to colony conditions which contribute to brood mortality include inbreeding and developmental anomalies. For female eggs, increased inbreeding results in a high brood death rate, often above 50% of the eggs laid by queens which have mated with their brothers (Woyke, 1962; Page, 1980). More commonly, both male and female eggs exhibit developmental anomalies before and after oviposition, or else fail to hatch (Fukuda and Sakagami, 1968; Fukuda and Ohtani, 1977).

ADULT DEVELOPMENT AND LONGEVITY

After emerging, the teneral bee completes its development during the next 8–10 days. The emerged bee is still soft, and the cuticle finishes hardening during the next 12–24 hr. The workers, in particular, have a soft, fuzzy appearance at this time until their hairs stiffen, and they cannot sting until the skeleton around the sting glands has hardened. Over the next few days internal development is completed, particularly glandular development and growth of fat bodies, but it is dependent on the young workers consuming sufficient pollen for protein. Insufficient pollen consumption early in life results in poor glandular development and a shorter life span (Maurizio, 1950; Haydak, 1970). The young workers get most of their protein from pollen they themselves remove from cells and eat. They begin consuming pollen within the first few hours after emergence and reach their maximum consumption level when about 5 days old (Morton, 1950; Hagerdorn and Moeller, 1967; Dietz, 1969; Haydak, 1970). There is also some feeding of brood food by nurse bees to young workers (Free, 1957a). This feeding period of emerged adults is important to the workers' development; within 5 days nitrogen content increases 93% in the head, 76% in the abdomen, and 37% in the thorax of workers (Haydak, 1934).

Drones and queens must also complete their development after emerging, but for them the principal source of protein is brood food fed to them by nurse bees. Just as in workers, their cuticle hardens within 24 hr, and development of internal organs and glands requires adequate nutrition. Unlike workers, however, these two castes must also develop their reproductive organs. In drones the seminal vesicles and mucus glands still must develop, and mature drone semen is not generally available until at least the twelfth day (Ruttner, 1966). In queens the ovaries remain small until egg laying begins about 2–4 days after mating (Gary, 1975).

Emergence weights of bees show as wide a range as development times and brood weights. For example, the ranges of average emergence weights for workers, drones, and queens are 81–151 mg, 196–225 mg, and 178–292 mg, respectively (according to 17 different studies reviewed by Jay, 1963a, and Lee and Winston, 1985a). Post-emergent adult bees show similar variability in weight; the range of mean weights per bee for samples of only 20 bees each was 81–140 mg (Park, 1925a; Mitchell, 1970). Factors affecting emergent worker weights include cell size, the number and age of nurse bees, colony population, availability of nectar and pollen, disease, and season (Jay, 1963a). Furthermore, the number of workers in newly colonized swarms influences worker weights; small swarms produce lighter workers, probably because of poor nutrition and nest homeostasis caused by a lack of workers to perform nurse, foraging, and thermoregulatory tasks (Lee and Winston, 1985a). Queen weights are influenced by similar factors as well as by the number and age of larvae being reared as queens at any one time. There is also a genetic component to worker weights; unengorged and post-emergent European and Africanized workers weigh on average 93 and 62 mg, respectively, although their average weights are considerably higher when engorged with nectar following swarming—130 and 93 mg (Otis, 1982). Although the biological significance of different body weights has not been well studied, in queens body weight seems to be related not to ovariole number (Eckert, 1934), but to colony performance and queen fecundity (Nelson and Gary, 1983). Worker body weights may influence the tasks performed by workers, with heavier workers tending to forage more than lighter workers (Kerr and Hebling, 1964), although this effect has not been confirmed by more recent studies (Nowogrodzki, 1984).

Once workers emerge, their life span can vary from only a few days to almost a year, depending primarily on seasonal factors, food availability, activities performed during their lifetime, and race. The general pattern in temperate climates is for workers to be short-lived in the summer; mean longevities of 15–38 days for summer bees have been recorded by numerous authors (reviewed by Ribbands, 1953; Fukuda and Sekiguchi, 1966; Michener, 1974; and Winston, Taylor, and Otis, 1983). Workers during the spring and fall have intermediate life spans, usually 30–60 days, whereas life spans in winter average about 140 days (Fukuda and Sekiguchi, 1966). However, workers in winter have been known to live 304 (Anderson, 1931) and even 320 (Farrar, 1949) days. The winter workers have well-developed hypopharyngeal glands and fat bodies from pollen consumption in the fall, characteristics that seem to contribute to their increased life span. Also, workers in winter are relatively inactive and have slower metabolic rates than summer bees (Corkins and Gilbert, 1932), which also increases their longevity. At any time of year, proper nutrition is necessary for workers to achieve their maximum potential longevities (Maurizio, 1950; Haydak, 1970).

The race of a worker is also important in determining its longevity, tropical bees being much shorter-lived than temperate races. For example, Africanized workers during the dry season in South America, which is equivalent to a temperate summer, survive on average only 12–18 days, many fewer days than the average for European bees (Winston, 1979c; Winston, Dropkin, and Taylor, 1981). The average survival time during the winter in Poland for African bees is only 90 days, again much shorter than for European races (Woyke, 1973a). Even among a single race of bees longevity has a strong genetic component. Kulincevic and Rothenbuhler (1982), for example, were able to select for long- and short-lived lines of Italian bees (*A. m. ligustica*).

The activities workers participate in also affect their life span. Longevity of the first Africanized workers to emerge in colonies after swarms are established is the lowest recorded for any honey bees, averaging only 12 days (Winston, 1979b). Their shortened life span is undoubtedly due to the massive amount of work which must be done to establish a nest, rear brood, and forage while colony populations are still low. Interestingly, workers of European races emerging in nests established by swarms show no such diminution of their life span, suggesting that the overall integration of work tasks and colony growth may be markedly different in tropical-evolved bees (Winston, Dropkin, and Taylor, 1981). Colony environment and sudden drops in worker population caused by such events as swarming, predation, nest damage, and disease may also influence life span (Winston and Katz, 1982; Winston and Fergusson, 1985).

Drones generally live an average of 21–32 days during the spring to mid-summer period, although mean life spans as short as 14 days and maximum mean longevities of 43 days have been noted (Jaycox, 1956; Drescher, 1969; Witherell, 1972; Fukuda and Ohtani, 1977). During late summer and autumn drones can survive up to 90 days, but as winter approaches they are usually expelled from the nest, and few or no drones survive the winter (Fukuda and Ohtani, 1977).

Queens are the most long-lived of the three honey bee castes, generally surviving for 1–3 years. In one study of the life history of unmanaged colonies, 79% of queens survived for 1 year, 26% for 2 years, and virtually no queens survived longer than 3 years (Seeley, 1978). These data agree with the recommendations for queen management in beekeeping situations that suggest requeening colonies about every 2 years. There are reports, however, of queens living much longer; Bozina (1961) found that up to 35% of queens in normal colonies lived 4–6 years, and he noted that three queens lived 8 years or more.

Nutrition

The three honey bee castes have somewhat different nutritional needs and feeding mechanisms to satisfy their food requirements, as do the larval and adult stages within each caste. But the starting materials for brood and adult bees, whether workers, queens, or drones, are the same: nectar and pollen. These two floral products provide all of the food necessary for larval growth and metamorphosis and for adult development and functions. Basically, nectar provides carbohydrates in the form of sugars, and pollen provides protein, lipids, vitamins, and minerals. Honey bees have evolved various mechanisms for processing nectar and pollen so that the food fed to each stage and caste are ideally suited to their needs.

NECTAR

Honey bees obtain most of their energy from carbohydrates in the form of plant-produced sugars, primarily from nectar produced in flowers but also occasionally from extrafloral nectaries or honeydew secreted by plant-feeding insects. Floral nectar is an aqueous plant secretion which contains anywhere from 5 to 80% sugars, and small quantities of nitrogenous compounds, minerals, organic acids, vitamins, lipids, pigments, and aromatic substances (White, 1975). Of all the latter substances, only ascorbic acid (vitamin C) is found in any appreciable quantity in nectar; the protein content of honey is usually less than 0.2% (White and Rudyj, 1978). Sucrose, glucose, and fructose are the main sugars found in nectar, and nectars can be broadly classified into three groups: (1) predominately or totally sucrose, (2) approximately equal proportions of sucrose, glucose, and fructose, and (3) predominately glucose and/or fructose. In addition to these three sugars, *a*-methyl glucoside, maltose, trehalose, and melezitose are of nutritional value to bees. Most other sugars neither taste sweet nor have nutritional value. Some sugars, such as mannose, galactose, and rhamnose, are either toxic to bees or cause reductions in their longevity (von Frisch, 1934, 1965).

The nectar collected by foraging workers can be fed to brood and adults directly, but it is more commonly processed into honey first (reviewed by Gary, 1975, and Maurizio, 1975). The nectar is carried back to the nest in the honey stomach and transferred to nest workers for processing. Enzymes from the hypopharyngeal glands are added to the nectar in the crop, specifically, diastase, invertase, and glucose oxidase. These enzymes break down the sugars into simple inverted forms, which are the most easily digestible by bees and also protect the stored honey from bacterial attack. The nectar is then evaporated on the worker's tongue and placed into cells for further evaporation by fanning; the water content generally is reduced to less than 18% to protect the nectar from yeasts. When the enzymatic activity and water evaporation are complete, the nectar is considered to be "ripened" and can now be called honey, which is sealed beneath a wax capping until required for feeding to larvae or adults. A worker larva requires about 142 mg of honey for development, and the annual honey requirements for a colony have been estimated at about 60–80 kg (Weipple, 1928; Rosov, 1944; Seeley, 1985a).

POLLEN

Pollen is the male germ plasm of plants, but in addition to being necessary for reproduction, many pollens have evolved so as to be attractive to and edible by bees. This is, of course, advantageous to both bees and plants, since bees require pollen for growth and development, and many plants require bees to transfer pollen between flowers. To the bee the most important constituent of pollen is protein. Pollens contain anywhere from 6 to 28% protein and are virtually the only source of protein naturally available to bees. Pollens also contain lipids (anywhere from 1 to 20%, but usually less than 5%), which are also important for bee nutrition. Another material in pollen, sterols, is present in minor but essential quantities. Most pollens contain less than 0.5% sterols, but these are required for bee metabolism since bees cannot synthesize cholesterol without the precursors obtained from pollen. Pollens also contain sugars, starches, vitamins, and minerals, all of which are important for bee nutrition. There is considerable variability in the nutritive value of pollens from different plants, partly because of the different amounts of protein in pollens; the importance of other compounds found in pollen is not as well understood (Parker, 1926; Vivino and Palmer, 1944; Maurizio, 1954, 1960; Stanley and Linskens, 1974).

Once pollen has been brought back to the colony by foragers, the workers treat it to prevent germination, begin the digestive process, and prepare it for long-term storage. A phytocidal acid is added when pollen is packed into the comb, which prevents germination and deleterious bacterial activity. The chemical nature of this substance is still not known, but it appears to be produced in the hypopharyngeal and/or mandibular glands (Maurizio, 1959; Chauvin and Lavie, 1956; Lavie, 1960; Pain and Maugenet, 1966). It may be related to 10-hydroxy-2-decenoic acid (Keularts and Linskens, 1968). Some preliminary digestion of pollen also takes place as a result of enzymes added by workers and possibly some beneficial bacterial action. For example, sucrose is largely inverted by worker-added invertase, and stored pollen has a high content of histamine and vitamin K, indicative of bacterial action. The enzymes added with honey when pollen is packed in the cells prevent anaerobic metabolism and fermentation, however, which contributes to the longevity of stored pollen in cells. When the pollen has been completely processed for storage it is often referred to as "bee bread," since it is now ready for ingestion and digestion by bees.

A process not yet well understood, pollen digestion provides challenges to both larvae and adults because of the hard and undigestible pollen wall components. When ingested, pollen is quickly passed through the honey stomach to the midgut, where digestion takes place. There appears to be very little mechanical breaking of pollen grains by either the mouthparts or the proventricular valve (Parker, 1926; Whitcomb and Wilson, 1929). Once in the midgut, the enzyme-secreting membranes tighten around the pollen bolus, at the same time protecting the midgut from the sharp spines and abrasive surfaces of the pollen wall (Barker and Lehner, 1972). The digestion of usable pollen nutrients takes place either through the pollen germination pores or through the pollen wall, which is ruptured by osmotic shock (Kroon, van Praegh, and Velthuis, 1974; Stanley and Linskens, 1974; Klungness and Peng, 1984). It generally takes 1–3 hr for a pollen mass to pass through the digestive system. The digestive process is similar in larvae and adults, although young larvae are not fed very much pollen directly. The amount of pollen required to rear a single worker larva has been estimated at 125–145 mg, containing about 30 mg of protein (Alfonsus, 1933; Rosov, 1944). A colony's annual requirements vary, but figures ranging from 15 to 55 kg have been reported (Eckert, 1942; Ribbands, 1953; Hirschfelder, 1951; Louveaux, 1958; Seeley, 1985a).

Feeding Behavior

WORKER LARVAE

Worker larvae are fed primarily brood food produced by the hypopharyngeal and mandibular glands of nurse bees, although some pollen is fed directly to larvae on the fourth or fifth day of larval development. The brood food consists of a clear component from the hypopharyngeal glands, which is presumably mixed with honey, digestive enzymes, and water, and a milky white component, which appears to be the mandibular gland secretions with an admixture of the hypopharyngeal gland secretions. Worker larvae are fed 20–40% from the white component and 60–80% from the clear component during the first 2 days of larval life. On the third day the amount of mandibular gland secretion fed to developing workers decreases, and the brood food originates mostly from the hypopharyngeal glands; a drop in diversity and quantity of protein fed to worker larvae occurs at this time (Shuel and Dixon, 1959; Patel, Haydak, and Gochnauer, 1960). Some pollen and honey are fed directly to larvae beginning on or after the third day, with the heaviest pollen feeding occurring on day 5 (Morton, 1950; Jung-Hoffman, 1966; Matsuka, Watabe, and Takeuchi, 1973; Michener, 1974). (The feeding of larvae by nurse bees is summarized in Chapter 6.)

The food requirements of larvae have never been precisely determined, and to date there is no artificial diet which can completely replace honey and pollen (for reviews, see Doull, 1977, and Chalmers, 1980). Nevertheless, some information on essential larval nutrition is available from experiments in which the amount of food or its composition have been manipulated. When worker larvae are underfed, dwarf adults are produced, and developmental failure is high if less than 65% of the normal amount of food is fed to larvae. Such underfeeding, and the resulting dwarf adults, can occur when nectar and pollen are scarce, colonies are diseased, brood is reared early in the spring, or colonies are low in worker population, all factors that reduce the quantity or quality of brood food (Jay, 1964b, and references cited therein). Adult bees can rear brood for a short time when fed a pure carbohydrate diet, but they must break down their own body tissues to produce larval food, and the emerging adults have a lower nitrogen content in their abdomens than in colonies with access to pollen (Haydak, 1935). The B vitamin pyridoxine and inositol are required for larval development, and when pyridoxine and cholesterol are added to a vitamin-free diet, normal development continues for at least four cycles of brood rearing (Haydak and Dietz, 1965, 1972; Anderson and Dietz, 1976). Vitamin C fed to adult workers results in a higher brood survival rate than in vitamin C deprived bees (Herbert, Vanderslice, and Higgs, 1985). Water is also essential for proper larval growth and development (Petit, 1963; Haydak, 1970).

Although lipids are not generally considered major nutrients for insects, some of the lipid components of brood food are essential for proper growth and development. Larvae fed lipid-deprived brood food gain weight and grow more rapidly than larvae fed normal brood food, but lipid-free diets result in an increased mortality rate at the prepupal stage and a failure to spin silk (Smith, 1960; Petit, 1963; Haydak, 1970). Thus, the removal of lipids reduces dietary components that limit growth to a normal rate and are essential for metamorphosis. Two of these components have been identified as the fatty acid 10-hydroxy-trans-2-decenoic acid (10-HDA), which is produced in the mandibular glands and makes up about 4% of the brood food (Butenandt and Rembold, 1957; Barker et al., 1959; Callow, Johnston, and Simpson, 1959), and sterols, mostly 24-methylene cholesterol, which makes up about 0.25% of brood food (Barbier and Bogdanovsky, 1961). The readdition of these components to lipid-free diets prevents accelerated larval growth and allows for normal pupation and subsequent adult emergence (Kinoshita and Shuel, 1975).

ADULT WORKERS

The basic nutrients of adult workers are honey and pollen, although some brood food is probably transmitted during food exchange between workers and used for nutrients. The nectar or honey eaten by workers provides sugars for energy; pollen cannot substitute as an energy source, and workers will die without an adequate supply of honey. Pollen is necessary for proper post-emergence glandular development and growth of internal structures during the first 8–10 days of a worker's life, but after that it is not essential unless older workers begin to produce brood food and feed larvae.

Most of the energy required by adult workers is obtained by lapping up honey from cells, although nectar and honey can be exchanged among workers. To initiate food exchange, a bee begging for food thrusts the tip of her tongue toward the mouth of another bee, and an offering bee opens her mandibles, pushes the tongue forward, and regurgitates from the honey stomach a drop of fluid, which is lapped up by the begging bee. Brood food also can be exchanged in this manner, although such food transfer occurs in fewer than 5% of trophallactic interactions (Free, 1957a; Korst and Velthuis, 1982).

The amount of sugar required by workers depends on their activity level; resting workers (*A. m. ligustica*) need only 0.7 mg sugar/hr, whereas those in flight require 11.5 mg/hr (Olaerts, 1956; Heinrich, 1979c). As mentioned previously, bees utilize only certain sugars and derive different amounts of energy from them.

Experiments in which bees are deprived of protein have demonstrated the importance of pollen to adult workers. Lack of protein during the first 8–10 days of adult life results in shorter life spans (Haydak, 1937a,b; Maurizio, 1950, and references cited therein; de Groot, 1951; Weiss, 1984) and poor development of the hypopharyngeal glands and fat bodies (de Groot, 1953; Maurizio, 1950, 1954; Back, 1956; Haydak and Dietz, 1965; Haydak, 1970; Anderson and Dietz, 1976). The type of pollen can be important; Maurizio (1960) classified pollens into groups based on their nutritive value, and Knox, Shimanuki, and Herbert (1971) showed reduced longevity in bees fed dandelion pollen. Of the protein required for proper growth, only ten of the amino acids are essential: arginine, histidine, lysine, tryptophan, phenylalanine, methionine, threonine, leucine, isoleucine, and valine (de Groot, 1953).

The requirements of adult bees for lipids, vitamins, and minerals have not been well studied. Vitamins were found to be necessary for hypopharyngeal gland development in two studies (Serian-Back, 1961; Standifer and Mills, 1977), but development of that gland and fat bodies were found to be normal on a vitamin-free diet in another study (Maurizio, 1954). Older workers show increased quantities of various minerals, which may be related to the tasks they perform, but this relationship has not been examined (Dietz, 1971). Adult workers seem to have lipid requirements similar to those of larvae, though in lesser amounts, but whether they need lipids in their diets beyond what is consumed as larvae has yet to be determined.

DRONE LARVAE

Relatively little is known about nutrition of drone larvae, although there appear to be some differences in the composition of the brood food given to developing drones and workers. Drones receive much more food than workers because of their larger size, and in addition their brood food has more diverse proteins than does worker brood food (Gontarski, 1954). Also, the food of older drone larvae contains more carbohydrates, riboflavin, and folic acid than that of younger larvae, and less thiamin, biotin, pantothenic acid, choline, pyridoxine, protein, fat, ash, and niacin. These changes are due in part to an increased quantity of honey and pollen in the diet of older larvae, but they also reflect variations in brood food composition (Haydak, 1957; Matsuka, Watabe, and Takeuchi, 1973).

Adult drones are fed entirely by workers for the first few days of their lives and then gradually begin to feed themselves from honey cells; after they are about a week old they feed themselves entirely. The young drones beg food from adult workers of any age but are only fed by young workers of brood food producing age. The food given to the emerged drones is a mixture of brood food, pollen, and honey, and in some cases regurgitated contents of the honey stomach (Free, 1957b; Ruttner, 1966; Haydak, 1970). The production of spermatozoa is not dependent on the amount of protein fed to young drones, but brood food and pollen feeding may influence longevity and mating ability, although this has not been examined directly. Older drones feed themselves exclusively from honey in the comb, which provides energy for mating flights. A resting drone requires only 1–3 mg sugar/hr, whereas flying drones consume 14 mg/hr, (Mindt, 1962), slightly more than workers.

QUEENS

The quantity and quality of food fed to female larvae determine whether those larvae will develop into workers or queens, as will be discussed in the next section on caste determination. Adult queens are fed by workers of brood food producing age and presumably receive mostly brood food, possibly with some additional honey (Allen, 1955, 1960; Haydak, 1970). Isolated queens can feed themselves on sugar candy and survive for many weeks (Weiss, 1967), but queens in colonies seldom, if ever, feed themselves. As might be expected, the extent of queen feeding is related to the queen's egg-laying rate; the number of feeds, duration of individual feeds, and total feeding time increase as colonies grow and the queen's egg-laying rate increases (Chauvin, 1956; Allen, 1960) (Fig. 4.5). A queen will lay as few as 2 or as many as 26 eggs between feedings, and is generally fed by only one worker between each period of egg laying, although up to five workers in succession have been seen to feed queens (Istomina-Tzvetkova, 1953; Allen, 1955).

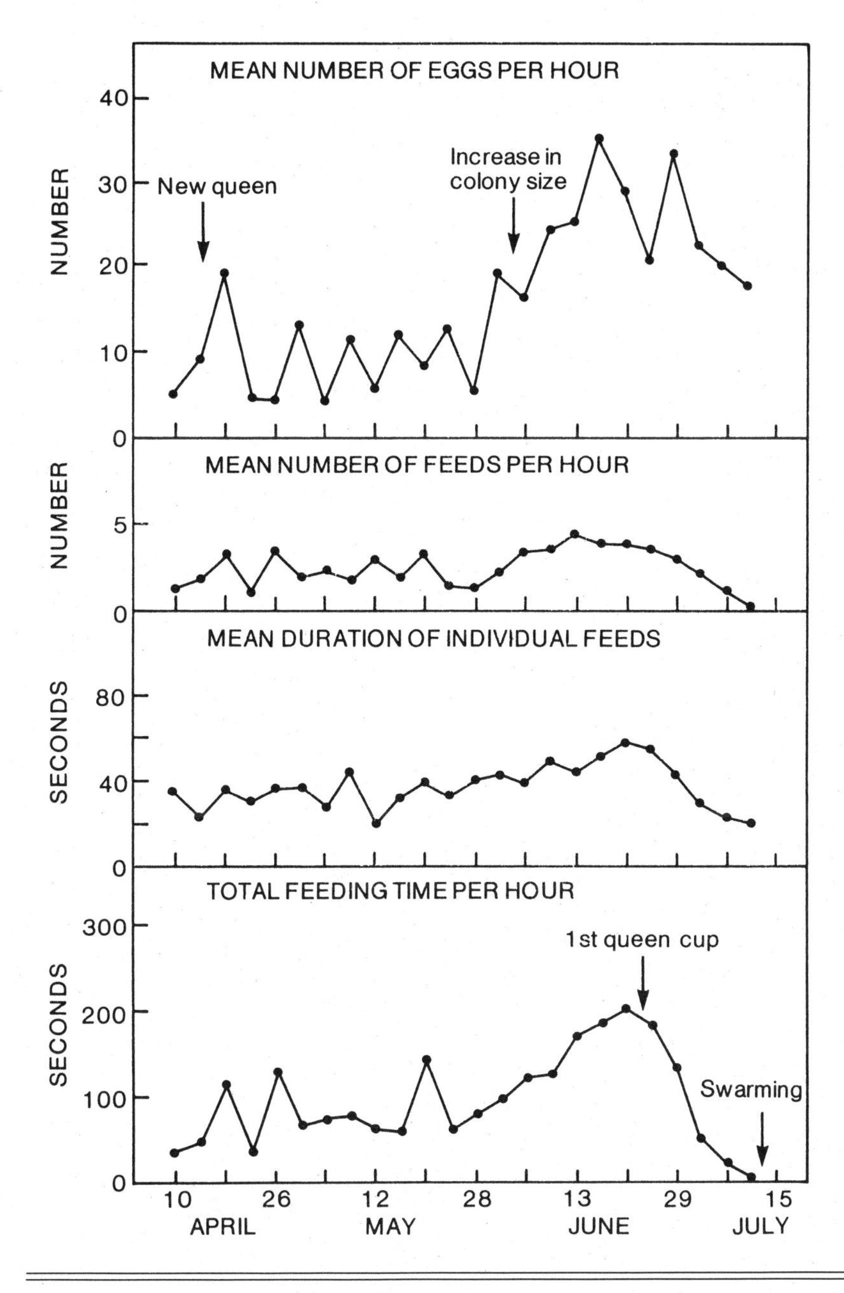

Fig. 4.5

Feeding and egg-laying rates of the queen. (Redrawn from Allen, 1960.)

Caste Determination

The mechanism of caste determination in honey bees can be expressed simply: unfertilized eggs develop into drones, and fertilized eggs develop into workers or queens, depending on the quality and quantity of the brood food fed to the female larvae (Fig. 4.6). But this simplification masks a very complex process, with many exceptions to the normal developmental pathways, one in which both genetics and nutrition have an important part. Much of what is known about caste determination has come from inducing bees to produce unusual adults, such as diploid drones, worker-laid males and in some cases parthenogenetic females, queens initiated with old larvae, and intercastes (individuals with both worker and queen characteristics). Considering the many things which can go wrong in caste determination, it is remarkable that colonies produce the intended caste as often as they do.

GENETICS

The primary differentiation in bee castes is male versus female. The reason for this differentiation involves the sex determination mechanism for all Hymenoptera and some other insects; haploid eggs develop into males, and diploid eggs become females. Thus, if sperm are not released when the queen lays an egg, that unfertilized egg has only the haploid number of chromosomes, or one set. If a sperm unites with the egg, the diploid number is restored, and a female is produced. By controlling the release of sperm prior to egg laying, the queen can determine whether that egg will mature into a male or female.

Since queens almost always lay unfertilized eggs in the larger drone cells and fertilized eggs in the smaller worker cells, they must be able to determine cell size prior to sperm release. According to one hypothesis explaining how queens determine cell size, since the worker cell is smaller than the drone cell, the pressure from the cell wall squeezes the sperm out of the spermatheca and the egg gets fertilized (Petrunkewitsch, 1901). This appears unlikely, however, because queens lay eggs properly in cells which have not yet been elongated, and the anatomy of the spermatheca is not consistent with this idea (Koeniger, 1970). A second hypothesis is that the queen determines cell size during her pre–egg-laying inspection. Prior to egg laying the queen places her head and forelegs into a cell, then withdraws her head, turns and curves her abdomen, and oviposits into the cell. Amputating all or part of the forelegs causes queens to lay many of their female eggs into drone cells, although drone eggs are not laid in worker cells. This suggests that fertilization of the egg is

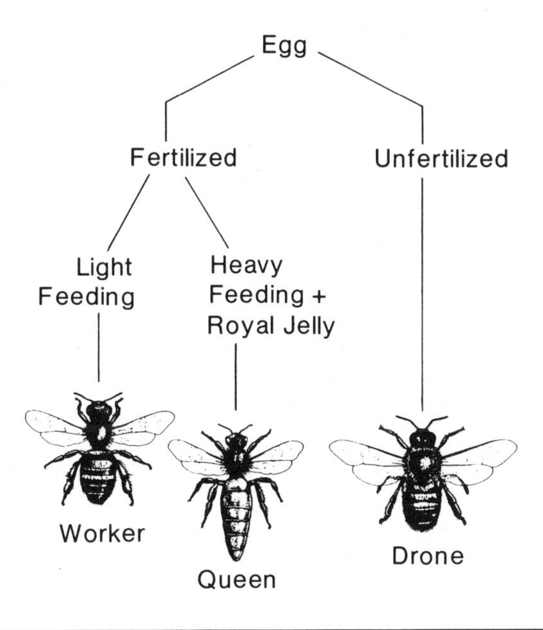

Fig. 4.6

Factors determining the differentiation of queen-laid eggs into workers, queens, or drones.

prevented by a stimulus obtained during inspection of drone cells with the forelegs (Koeniger, 1969, 1970). An additional mechanism for cell measurement might also involve the abdominal angle upon oviposition in the two different-sized cells.

Queens occasionally lay fertilized eggs in worker cells, which can potentially develop into males called diploid drones. An explanation of this phenomenon was first proposed by Mackensen (1951), who suggested that sex in fertilized honey bee brood is determined at a single gene locus with multiple alleles. Estimates of the number of alleles range from 6 to 18, and individuals homozygous at this locus become diploid males (reviewed by Page, 1980). The diploid drone condition is invariably lethal, however, since diploid male larvae are eaten by workers within 72 hr of eclosion, possibly because of a substance secreted by diploid drone larvae which causes cannibalism (Woyke, 1963). Normal male eggs, being haploid, are hemizygous, and so do not express this characteristic. By rearing diploid drones artificially, Woyke (1969, 1973b) was able to show that diploid males, when reared to adults, have small testes and produce little semen, rendering them functionally infertile. Thus, by eating young diploid drone larvae the workers prevent wastage of food and space to rear nonfunctional individuals.

Under certain conditions workers also can lay eggs, which usually develop into drones. Although a significant amount of egg laying by workers does not occur in queenright colonies, workers with developed ovaries are found in most normal colonies. The percentages of these workers vary between 7 and 45 in normal colonies, and between 20 and 70 in colonies about to swarm (Tuenin, 1926; Perepelova, 1928a; Koptev, 1957), although they probably lay few or no eggs unless the colony becomes queenless. When queens are lost and colonies fail to rear a new queen, the workers with enlarged ovaries begin to lay eggs within a few weeks, although these eggs generally can only develop into drones since the workers are not able to mate. The eggs of laying workers are smaller than queen-laid eggs and are placed on the sides rather than at the bottoms of cells, since the worker abdomens do not quite reach the bottoms of cells when inserted for egg laying. Many worker-laid eggs are often found in a single cell, although each worker only lays one egg per cell. These eggs can develop into fertile drones, although the adults are usually smaller and possibly not as successful at mating as queen-laid drones (reviewed by Ribbands, 1953; Gary, 1975).

Finally, females can sometimes develop parthenogenetically from unfertilized eggs laid by either queens or workers. Virgin queens not permitted to mate lay some eggs which develop into workers; only 1% or less of all eggs are laid this way, but these eggs can be reared into queens through two generations (Mackensen, 1943). Laying workers of the Cape bee *A. m. capensis* regularly lay eggs which develop into workers or queens, and laying workers of other races may do the same, although not as frequently (Anderson, 1963). The mechanism for parthenogenetic development is automictic; that is, following meiosis recombinations of chromosomes occur which restore the diploid number of chromosomes and result in female development. Such recombination must result in heterozygous alleles at the sex locus or, as for fertilized eggs, diploid drones will be produced (Tucker, 1958; Verma and Ruttner, 1983).

NUTRITION AND HORMONES

It was realized early on that a fertilized egg has the capability of developing into a worker or a queen, depending on the type of cell it is laid in and some aspect of nutrition (see Fig. 4.6). An egg laid in a worker cell can be moved into a queen cell and, under the proper conditions, will develop into a queen; conversely, an egg transplanted from a queen to a worker cell will be reared as a worker. However, while the type of cell may be an important stimulus that provides information to workers about how to rear its occupant, cell type alone does not explain the mechanism which determines

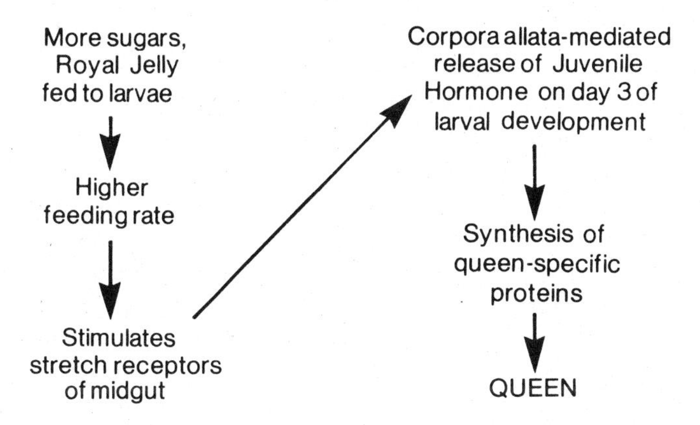

Fig. 4.7

The nutritional and hormonal factors which induce larval differentiation into queens.

whether a worker or a queen will be reared. Rather, the quantity and quality of the food given to a developing larva determines its caste, and these nutritional factors act through the larval hormonal system (Fig. 4.7). For recent reviews of female caste determination in honey bees, see Weaver (1966), Michener (1974), Goewie (1978), Beetsma (1979), and de Wilde and Beetsma (1982).

The food of queen larvae is referred to as "royal jelly," and it differs from worker food in containing more mandibular gland secretions and in the quantity that is fed to the brood. The food of worker larvae contains a ratio of 2:9:3 of white to clear to yellow components (averaged over the 5 days of larval feeding). The white component contains mandibular gland secretions, the clear material originates from the hypopharyngeal glands, and the yellow component is mostly pollen. In contrast, queen larvae receive mostly white food during the first 3 days, and a 1:1 ratio of white to clear during the last 2 days of feeding (Jung-Hoffman, 1967). Thus, queen larvae receive a much higher proportion of mandibular gland secretion than do worker larvae. Further evidence for the predominance of mandibular gland secretions in royal jelly has been provided by chemical analyses which show that royal jelly contains up to 10 times more pantothenic acid and 18 times more biopterin than the material fed to worker larvae (Haydak and Vivino, 1950; Lingens and Rembold, 1959; Hanser and Rembold, 1960). The mandibular glands of queenless workers contain 8 times the pantothenic acid and 115 times the biopterin found in queenright workers, while the hypopharyngeal glands show no differences between queenright and queenless workers, confirming that these substances are of mandibular gland origin (Hanser and Rembold, 1964).

But the quality of royal jelly does not totally explain differentiation between queen and worker; queen larvae must also consume more food than workers in order to develop into adult queens. The growth and respiratory rates of queens and workers are similar for the first 2–3 days of larval feeding but then accelerate in queens during the last few days (Melampy and Willis, 1939; Osanai and Rembold, 1968; Dietz and Lambremont, 1970). The difference in the amount fed to queen and worker larvae is easily observed by examining cells of the two larval types; queens literally swim in a sea of brood food, whereas worker larvae are found with much less food in their cells. The principal feeding stimulant for queen larvae is sugar. Royal jelly has 34% sugar content, and worker food has only 12% for the first 3 days, although it increases to 47% during the fourth and fifth day owing to the addition of honey (Shuel and Dixon, 1959). The nature of the sugars differs as well. In the diet of queen larvae glucose is the main sugar component throughout the larval period, whereas for workers glucose is predominant during the early larval stages and fructose during the later stages (Brouwers, 1984). Furthermore, the addition of sugar to worker larval food induces more queens and intercastes to develop (Asencot and Lensky, 1976). The phagostimulant quality of sugar has been investigated by Goewie (1978), who demonstrated that larvae are capable of perceiving sugar content of food and respond more to royal jelly than to worker brood food. Also, worker larvae reared in drone cells are heavier than those reared in worker cells and have larger acid glands and a higher number of ovarioles in their ovaries, presumably because of heavier feeding (Nogueira and Goncalves, 1982). The behavior of nurse bees also demonstrates the extra feeding given to queens; queen larvae are visited ten times as often as worker larvae (Lindauer, 1952). Nurse bees probably do not specialize in feeding worker or queen larvae but seem to vary the proportions of their mandibular and hypopharyngeal gland secretions according to the type of larva they are visiting (Free, 1960a).

Although worker and queen development is clearly based on nutritional factors, the effects of differential feeding are not expressed until larvae are at least 3 days old. Eggs or larvae transferred from worker to queen cells or vice versa are totipotent; that is, for the first 3 days they can change their development to correspond to the type of cell they are placed in (Becker, 1925; Lukoschus, 1956). Larvae that are transferred between 3 and 4 days of age develop into intercastes, producing either queenlike workers or workerlike queens. For ex-

Table 4.1 Mean characteristics of honey bee queens produced by transferring eggs or larvae from worker cells to queen cells at various stages

		Larval age			
Stage of transfer	Egg	1 day	2 days	3 days	4 days
Weight (mg)	209	189	172	147	119
No. of ovarioles	317	308	292	272	224
Spermathecal diameter (mm)	1.31	1.28	1.21	1.16	1.03
Spermathecal volume (mm^3)	1.18	1.09	0.94	0.82	0.59

Source: Woyke, 1971, as interpreted by Michener, 1974.

ample, larvae transferred from worker to queen cells at this age may become queenlike adults but possess such worker characteristics as pollen baskets, a barbed sting, and worker mandibles (Weaver, 1957; Asencot and Lensky, 1984). Queen characteristics are also reduced in size and number; intercaste queens reared from larvae 3–4 days old have fewer ovarioles, smaller spermathecae, and weigh less than individuals whose rearing as queens began as younger larvae (Woyke, 1971) (Table 4.1). Caste is fully determined by the fourth day, and few larvae survive after transfer between cell types at this age. Those that do survive become almost typical workers or workers with some queen characteristics.

The mechanism for the manifestation of nutritional influences on caste determination has been thought to involve the endocrine system (Haydak, 1943; Lukoschus, 1955; Shuel and Dixon, 1960; Dixon and Shuel, 1963; Canetti, Shuel, and Dixon, 1964), but it has only been during the last decade that this suggestion has been confirmed and documented. The most well-studied hormone involved in caste differentiation in honey bees is juvenile hormone (JH), which is known in most advanced insects to regulate growth, development, metamorphosis, and many aspects of behavior. JH is secreted by the corpora allata, a relatively large globular organ found on the sides of the esophagus of both larvae and adults. The corpora allata is connected by nerves to two other endocrine organs, the corpora cardiaca and the neurosecretory cells of the brain (Snodgrass, 1956; Breed, 1983a), and is thought to act through its influence on molting hormone secretion from the corpora cardiaca and as yet unidentified substances from the neurosecretory cells. In larvae changes in JH levels induce molts between larval to pupal and pupal to adult stages, and in adults JH is involved in regulation of age-dependent polyethism and glandular development (reviewed by Robinson, 1985; see also Chapter 6).

Juvenile hormone also effects differentiation between queen and worker in honey bees. Research elucidating the role of JH has taken two approaches: (1) surgically removing or adding the JH-producing corpora allata of larvae, and (2) measuring and manipulating the amount of JH which larvae have been exposed to. These techniques involve delicate microsurgery, precise timing of treatments, and meticulous measurements of the results of experimental manipulations. In experiments in which the corpora allata of queen larvae 3–4 days old were implanted into worker larvae 4 days old, the worker larvae that emerged had more ovarioles than sham-operated and control worker larvae (Ber-Lin Chai and Shuel, 1970). In more extensive experiments Wirtz (1973) was able to demonstrate that the pattern of corpora allata activity in queen larvae suggested a higher rate of JH synthesis from the third to the fifth day of development than that found in worker larvae. He extended these studies to demonstrate that the JH titer in queen larvae 72 hr old was more than ten times that of comparable worker larvae and remained at the same high level until the beginning of the prepupal stage, when it diminished to the level found in worker prepupae (see also Lensky, Baehr, and Porcheron, 1978). The conclusion drawn from these studies is that the level of JH in queen larvae rises steeply on the third day of development and remains high throughout the larval feeding stages. The high titer of JH in queen larvae of these ages induces development into queens rather than workers.

The importance of JH in queen development and its mode of action have been further examined by applying JH to developing larvae under different conditions and monitoring its effects. If JH is involved in queen-worker differentiation, then addition of JH to worker larvae should induce queens or queenlike intercastes, but only at the proper ages. Also, a combination of feeding some of the royal jelly constituents along with JH application should provide the best induction of queenliness. When JH is applied topically to worker larvae and only worker brood food is used for feeding, individuals with queen characteristics develop (Wirtz and Beetsma, 1972; Wirtz, 1973). In similar experiments in which fructose and glucose are added to worker food and JH also applied, intercastes or queens are reared. Higher levels of the sugars and of JH result in queens, and lower levels produce intercastes or workers (Asencot and Lensky, 1976, 1984). In a more subtle test Naisse (cited in Beetsma, 1979) was able to show that JH application prevents the degeneration of ovarioles in worker larvae as early as the second day of larval development. Similar effects are also found in experiments using JH analogues, substances with different molecular structures than JH that nevertheless have similar effects on insects (Beetsma, 1979).

In summary, workers or queens can potentially develop from the same egg, while nutrition mediated by juvenile hormones determines which caste will develop (see Fig. 4.7). Individuals destined to become queens are fed royal jelly, which is particularly rich in mandibular gland products and sugars that are phagostimulants. The quality of this food as well as the quantity consumed induce high levels of JH during the critical third day of development, resulting in differentiation into queens. Worker larvae are fed more hypopharyngeal gland secretions during the first few days of the larval stage, and more honey and pollen during the last few days, resulting in lower JH levels on days 3–5 and differentiation into workers.

The major challenge for the next generation of researchers will be establishing the role of the brain and neural stimuli in mediating the interface between nutrition and hormones. A reasonable hypothesis has been developed by Goewie (1978) and Asencot and Lensky (1984), which proposes that the level of sugars in food controls the rate of food intake, and further that higher rates cause midgut extension, excitement of stretch receptors of the gut, and subsequent stimulation of the corpora allata to synthesize and release JH. The higher levels of JH in turn induce queen-specific proteins and enzymes which act on developing tissue to produce queens; lower levels of JH induce worker-specific development. The target tissue, however, must be competent to be induced by JH, and this occurs about the third day of development (Shuel and Dixon, 1973; Shuel, Dixon, and Kinoshita, 1978). The testing and elaboration of this hypothesis will undoubtedly be of major research interest in the next few years.

5

Nest Architecture

Most of our contacts with honey bees are outside, where we see workers flying near flowers and occasionally going in and out of a hole in a tree, stump, or barn wall. For bees, however, these forays to the outside world are only a small, although important, part of their existence. A typical worker bee spends the first 15–20 days of her life entirely inside the colony's nest, and even when she begins to forage, she spends only a few hours a day outdoors.

The nest, almost citylike in the diversity of its functions, is constructed inside a log cavity with a small, easily defended entrance. Inside, the bees construct wax combs with hexagonal cells in which they rear brood and store nectar, honey, and pollen (Fig. 5.1). Bees patrol the nest on the comb surface, performing or following the dances of foraging bees, passing food to each other, fanning to cool the nest and evaporate water from nectar, and clustering for warmth, among other activities. The nest itself is exquisitely designed for all these functions, and our gradually evolving understanding of the relationship between nest architecture and honey bee physiology, behavior, and ecology has been one of the most exciting aspects of contemporary bee research.

Choosing a Nest Site

Honey bees select a new colony site as the last stage of swarming, or colony reproduction. Colonies generally swarm in late spring, when the old colony has an excess of workers and has become overcrowded. At that time a majority of the workers leave the nest with a queen and form a cluster, usually under an overhanging limb or in a snarl of branches. The swarm then faces a critical problem; it must quickly find a new nest site before the workers run out of honey carried in their honey stomachs or the swarm population will begin to dwindle as workers die. The swarm also has to choose a site in which the new colony can survive and grow for many years.

Our understanding of how swarms select nest sites has come from

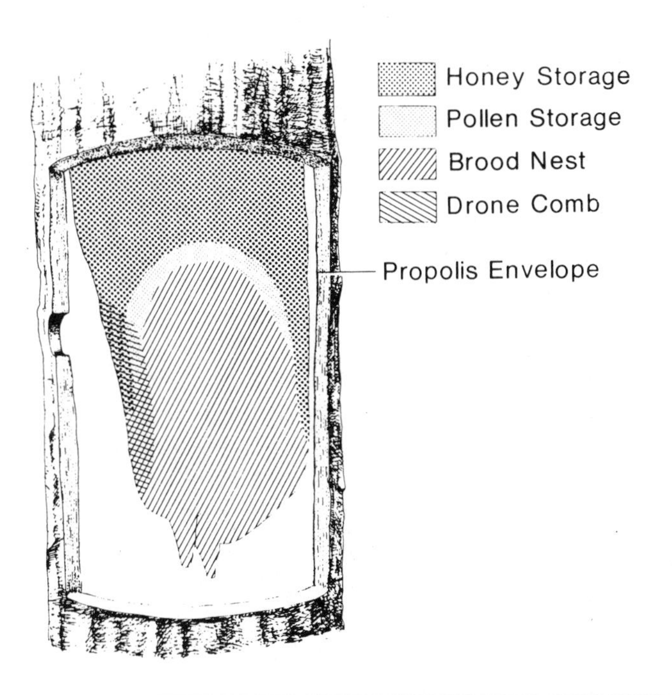

Fig. 5.1

A typical honey bee nest inside a cavity, showing the usual arrangement of cells for honey and pollen storage, worker brood rearing, and drone comb. (Redrawn from Seeley and Morse, 1976. Used by permission of MASSON S.A., Paris.)

three different types of research. First, many hours have been spent observing worker behavior in swarms and at potential nests in order to understand the selection process by which sites are evaluated. Second, feral nests have been dissected and every possible nest characteristic measured, a laborious task requiring a combination of lumbering skills and meticulous measurements. Finally, natural and artificial swarms have been given choices among cavities of various sizes and shapes to examine further how swarms choose nest sites. From these studies, which have involved numerous laboratories around the world for the last 40 years, we now have a fairly complete picture of the nest selection process and of what characteristics are important for colony survival.

The first studies examining how honey bees choose nests were performed in Germany by Martin Lindauer and colleagues during the late 1940's and early 1950's (Lindauer, 1951, 1955; von Frisch, 1967a). They realized that a swarm must be faced with numerous potential sites to scout and choose from, and the workers must reach a consensus as to the preferred site before the swarm lifts off and moves to it. This selection process can take anywhere from a few hours to a week or more but generally takes 1–2 days. By observing worker behavior during this clustering period, Lindauer was able to determine some nest selection behavioral patterns.

Once a swarm has settled into its interim clustering site, scouts leave the swarm cluster almost immediately and begin to search for appropriate nests. In many cases these scouts may begin looking for nests as much as 3 days before the swarm leaves the old nest. When a scout has found a potential cavity, she spends a considerable amount of time examining it. During the first phase of examination, the scout alternates brief inspections of the nest interior with short examination periods outside the nest cavity (Lindauer, 1955; Seeley, 1977). Outside behaviors consist of extensive crawling around the outer surface of the potential nest as well as slow hovering flights at increasing distances from the nest. Interior inspections involve rapid walking about the cavity's inner surfaces, interspersed with brief flights from one area of the nest to another. Scouts may groom each other, exchange nectar, and even scent at the entrance by exposing the Nasonov gland and giving off attractant chemicals, which are dispersed by wing fanning. Sporadic returns to the site throughout the day by the scout enable her to evaluate the site under different conditions.

When a scout finds a reasonable site, she returns to the swarm and does dances similar to those used by returning foragers to communicate the location of nectar and pollen sources. Nest scouts, however, rarely bring back nectar or pollen, and they are only communicating potential nest locations. These nest dances can last for 15–30 min, much longer than the dances associated with foraging, which last 1–2 min. Workers on the face of the swarm can "read" the location and quality of the nest sites by the tempo, angle, and duration of the scout's dances and are recruited by these dances to investigate the new sites. At first, many different dances can be seen on the swarm face, indicating that numerous cavities are being examined (Fig. 5.2). The number of dances to different sites gradually decreases as the workers reject unsuitable locations and begin to dance with great enthusiasm for the better sites. The swarm finally reaches a consensus when most of the scouts are dancing to the same location. At that point the scout bees perform a buzzing run, which causes the cluster to break up and take to the air. The workers then move with the queen to the new nest. "Guide" bees, which may be the scouts, seem to lead the swarm using a combination of rapid flight in the direction of the nest and chemical cues, although by the time the

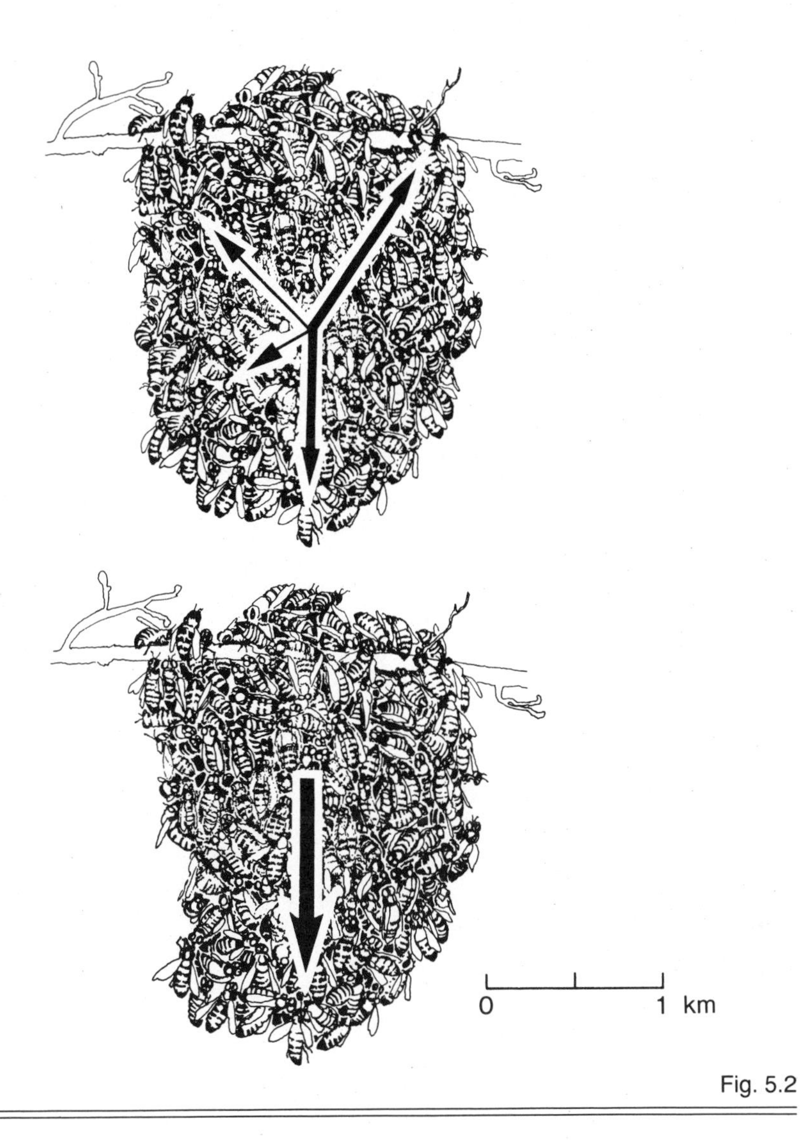

Fig. 5.2

The distance and direction of typical dances performed by scouts reporting on nest site locations. Each arrow represents dances to one nest site; the length of the arrow indicates distance to the potential nest, the angle of the arrow indicates the direction, and the thickness of the arrow indicates the relative number of scouts dancing to each site. The top figure shows dances to four sites before a consensus is reached, and the bottom figure shows a dance to one site just before the swarm moves to its new nest.

swarm lifts off many of the workers have read the dances and presumably know the new nest's location.

In one interesting example described by Lindauer the scouts could not come to an agreement and continued to dance vigorously to two different locations, one 1400 m to the WNW and the other 900 m to the NNE. The swarm eventually lifted off and split, and the two groups began to move to different sites. This confusing situation was resolved when the swarm reclustered about 30 m from its initial resting place, and the dances began again. After about 3 hr the dances to the NNE were becoming less frequent, and by nightfall the only dances to be seen were those to the site WNW.

Nest Site Characteristics

What are the characteristics that determine the quality of a nest site? One factor is the distance from the parent nest. A swarm has two opposing considerations which determine the optimal distance separating the new nest site from the parental colony. It is advantageous for a swarm to leave the foraging range of its parent nest so as to reduce competition for nectar and pollen, but long-distance flights are energetically expensive. To examine this conflict, studies have been conducted in which empty bait hives were placed at various distances from artificial swarms, and the distance to the hive chosen was noted (Fig. 5.3). Most studies of swarm movement have suggested that the optimal compromise between these opposing selection pressures is for swarms to move 500–600 m from the parent nest (Lindauer, 1955; Seeley and Morse, 1977). Of a total of 35 swarms, only 1 colonized a hive nearer than 300 m from the interim swarm clustering site, and only 5 traveled more than 1600 m. However, in some studies swarms preferred closer sites, with boxes 50–200 m away preferred over sites 400–800 m away (Jaycox and Parise, 1980, 1981).

These contradictory results may actually reflect differences between bee races or local resource conditions that could affect distance preferences for new nest sites. In experiments similar to those described above swarms of two different races were given choices of nesting sites from 0 to 1000 m away. The Mediterranean-evolved Italian bees (*A. m. ligustica*) preferred hives a mean of 155 m away, whereas the German bees (*A. m. carnica*), which evolved in a colder region, preferred nest boxes 690 m from the swarm clustering site. Italian scouts were never seen at boxes more than 350 m from the swarm, whereas German scouts rarely examined sites less than 300 m or more than 850 m away (Gould, 1982). Gould's interpretation of these results is that the more winter-adapted German bees need to

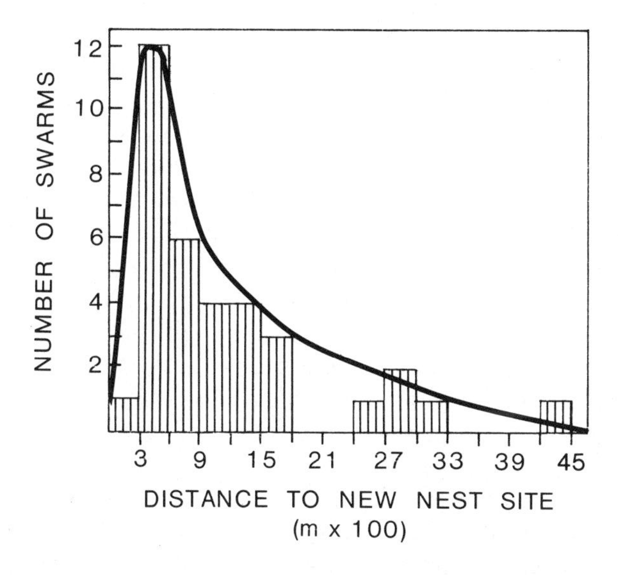

Fig. 5.3

Distances moved by 35 swarms to new nest sites. (From data in Lindauer, 1955, and Seeley and Morse, 1977.)

collect and store a much larger amount of honey to survive the winter than the Italian bees, and thus *A. m. carnica* swarms move further to minimize competition for nectar with the parent nest. This idea is supported by Gould's observations that the German swarms prefer larger cavities and use dance dialects which are more precise at greater distances than dances used by Italian bees. It is also possible that local resource conditions are perceived by workers, and the same race could show different distance preferences under differing resource conditions. For example, there is some evidence that swarms of tropical-evolved honey bee races migrate hundreds of kilometers during dearth seasons but only move a few hundred or thousand meters when strong honey flows are available (Fletcher, 1978a; personal observations).

Another important characteristic of nest sites is cavity volume. A swarm in a temperate climate must avoid cavities which are too small, since these would not have sufficient room to store the honey needed to survive winters and would be difficult to cool in the summers. Conversely, colonies might experience difficulty maintaining temperature in large cavities during winters, and oversize nest sites may also be difficult to defend. Evidence from nest dissections has shown that average nest volume is about 40 L, with most nest cavities in the range of 20–100 L (Wadey, 1948; Seeley and Morse, 1976). This average size is about equal to one standard Langstroth bee box or super, which is 42 L. However, occupied cavities as small as 12 L and as large as 450 L have been found (Seeley and Morse, 1976), and one 630-L nest with 200 kg of honey was reported by Percival (1954). Experiments also have been done which give swarms choices of nest volumes, and the results have shown that swarms generally prefer nests of 40–80 L in volume (Marchand, 1967; Seeley and Morse, 1977; Jaycox and Parise, 1981; Gould, 1982). From all of these experiments and nest dissections we can conclude that swarms in colder temperate areas prefer cavities about 40 L in volume, although they commonly occupy cavities 20–80 L in size, and colonies have been occasionally found in smaller or larger nests.

Different bee races also have different preferences for cavity volumes; temperate-evolved races are found in much larger cavities than subtropical- or tropical-evolved races, and tropical races are frequently found nesting outside cavities (Fig. 5.4). For example, the subtropical Italian bees prefer artificial boxes 30 L in size, whereas the more hardy German bees prefer 60-L boxes (Gould, 1982). In South America feral nest cavities of the recently introduced African bees have an average volume of 22 L, with nests as small as 7 L (Winston, Taylor, and Otis, 1983). Even when these tropical bees are found in larger cavities, they rarely occupy more than 25–30 L of their nest (Winston and Taylor, 1980), in contrast with temperate bees, which generally fill most or all of their nest cavity (Seeley and Morse, 1976).

Even more dramatic are the external nests characteristic of tropical honey bees (Fig. 5.4). These can be found under branches and overhanging rocks or in buildings and occupy volumes similar to cavity nests, about 20–30 L. In some areas most of the wild nests are found outside cavities, although this kind of sample is biased because external nests are easier to find. How such sites are chosen by swarms is not known, but they may be interim clustering sites, at which workers began constructing comb and the swarm remained. Since such nests are usually not found near an established colony, it appears that swarms move at least once before beginning nest construction outdoors, but it is not clear whether scouts indicate a particular site or whether the swarm moves cross-country without a specific site, settles in a likely location, and begins comb building. The latter is more likely, since many cases have been observed in which tropical swarms lifted off and moved without scouts dancing to indicate a potential nest or clustering site.

Why might small nests and nests external to cavities be common for tropical bees and not for temperate ones? First, tropical bees do not need to store large amounts of honey to survive the winter and thus do not require the large colonies typical of temperate bees. Sec-

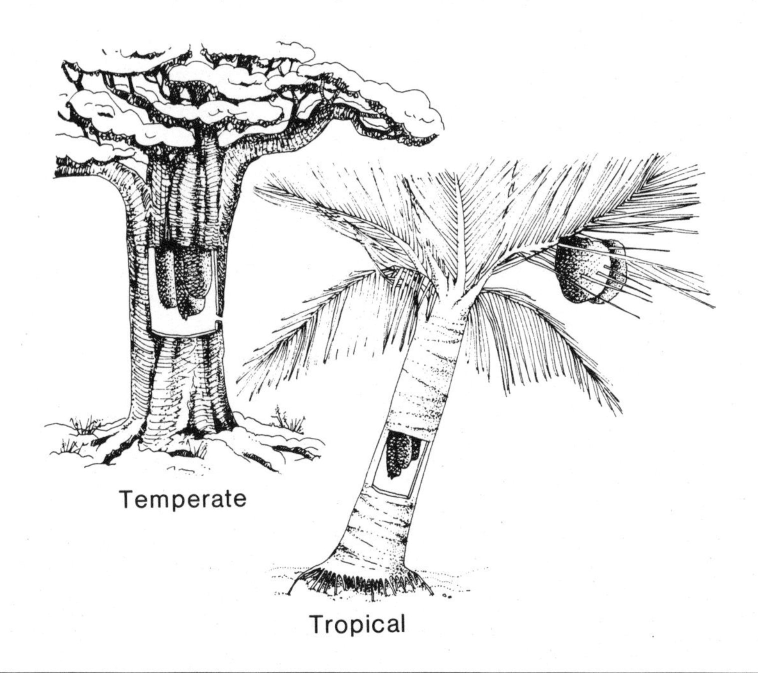

Fig. 5.4

Typical nesting situations for temperate- and tropical-evolved honey bees. The temperate races usually nest inside cavities, whereas the tropical races nest either inside cavities or externally, and their nests tend to be smaller than those of temperate races.

ond, a large worker population is not necessary to maintain colony temperature during the winter in tropical areas. Finally, predation on honey bees is much more intense in tropical habitats, and nests in small cavities are more easily defended. Also, if a predator succeeds in destroying a nest, much less is lost by the colony in a small nest, and after such a predator attack the remaining workers and the queen can leave the site and begin a new nest elsewhere (Winston, Taylor, and Otis, 1983). Thus, nest site choices seem to reflect adaptive differences between temperate- and tropical-evolved bee races.

Many other nest characteristics, in addition to volume, may be considered by scouts inspecting cavities. The following observations of temperate-evolved bees in the northeastern United States have come from dissections of feral nests and experiments giving swarms choices of different colony characteristics (Seeley and Morse, 1976, 1978; Avitabile, Stafstrom, and Donovan, 1978; Morse and Seeley, 1978).

Height above ground. In one study most entrances of feral nests were found more than 3 m from the ground (Avitabile, Stafstrom, and Donovan, 1978), while a second study found most entrances below 3 m (Seeley and Morse, 1976). The difference between these studies simply might reflect the distribution of available nests; in a choice experiment, however, six of six swarms preferred a nest box 5 m high over a 1-m-high box (Morse and Seeley, 1978).

Exposure and visibility. More visible nest boxes are preferred by swarms, but sites heavily exposed to wind, sun, and rain are not generally accepted.

Entrance size. Most entrances are found in knotholes, tree cracks, or holes between roots and are typically 10–20 cm in area. Colonies generally have only one entrance, although up to five have been found. The entrance size probably represents a compromise between the advantages of a larger entrance, in terms of summer ventilation and forager trips, and the benefits of a smaller entrance, in terms of winter heat retention and defense.

Entrance position. Entrances are usually located below the area where the bees build their comb inside the cavity, although swarms accept other entrance positions.

Entrance orientation. Feral nests and nest box occupation studies performed in the northern hemisphere both indicate that south-facing entrances are preferred. Cavities with a more southerly exposure get more sunlight and may be drier than those facing other directions.

Cavity shape. Feral nests are generally elongate and cylindrical, consistent with a tree's shape. However, choice experiments have shown no shape preferences, suggesting that the results from feral nest dissections are due to the shape of available cavities.

Cavity dryness and draftiness. Cavities with wet floors and drafty cavities are accepted by swarms. Workers can plug drafty and damp cavities with plant resins, and swarms are likely to reject only extremely wet or drafty cavities.

In summary, scouts investigate nest sites extensively and communicate nest location and quality to swarms by the vigor and orientation of their dances. In temperate climates swarms prefer nest sites away from the parent colony, about 40 L in volume, with a south-facing entrance 20–40 cm in area, and about 3 m above ground. Bee races from more tropical habitats colonize smaller cavities and may even construct external nests. Once a swarm has chosen its new site, the workers move to it with their queen and begin nest construction.

Comb Construction

Once the swarm has arrived at its final nest site, comb construction begins immediately. Many of the workers in the swarm have already begun producing beeswax for comb building, which can be seen as wax flakes protruding ventrally from between the abdominal segments. Rapid comb construction is essential to the swarm, since none of a colony's brood rearing or honey and pollen storage can be done without the comb. Over 90% of a feral colony's comb building is completed within 45 days of nest colonization, indicating the priority which the workers give to comb construction (Lee and Winston, 1985b).

The comb itself is one of the marvels of animal architecture. It consists of a regular back-to-back array of hexagonal cells, arranged in parallel series, each comb a precise distance from its neighbor. There are two types of hexagonal cells which make up the comb; the smaller cells are used for rearing worker brood, and drones are reared in the larger ones (see Fig. 1.1). Both types of cells can also be used to store honey, pollen, and occasionally water for short periods. A third type of elongate, conical cell, which hangs from comb edges, is constructed for queen rearing, but these are found only when colonies are swarming or superseding a failing queen. Only 10–20 of these cells are generally built, and the workers tear them down after a queen has emerged.

The hexagonal shape of cells is common among cell-building social insects, and there is a sound architectural reason for this style. Round, octagonal, or pentagonal cell arrangements leave empty spaces between cells, and triangles or squares have a greater circumference than hexagons (Fig. 5.5). Thus, the greatest number of cells per area can be arranged in comb using the hexagonal shape. By staggering each side of the comb slightly from the opposing side, the bees can construct cells back-to-back, further maximizing the number of cells per area. Unlike most other social insects, honey bees build their cells horizontally rather than hanging vertically, although they are angled up at about 13° from base to opening to prevent honey from running out (von Frisch, 1974) (Fig. 5.6).

In Italian (*A. m. ligustica*) and other European bee races, the worker cells are generally 5.2–5.4 mm in diameter, or 857 cells/sq dm, whereas the drone cells are 6.2–6.4 mm, 520 cells/sq dm (von Frisch, 1974; Dadant, 1975). Cell size, however, can vary with both bee race (Alber, 1956) and colony age. For example, African-evolved *A. mellifera adansonii* construct worker and drone cells 4.8–4.9 mm and 6.0–6.3 mm in diameter, respectively, and rear smaller bees from those cells (Smith, 1961). Construction of new comb also results in more precise cell structure than that found in older colonies. Newly colonized swarms of Italian bees constructed worker cells of a precise 5.2 ± 0.05 mm diameter (Lee and Winston, 1985b), whereas cells in older colonies can be highly irregular because of distortion by heavy loads of honey (Seeley and Morse, 1976). There is also some evidence

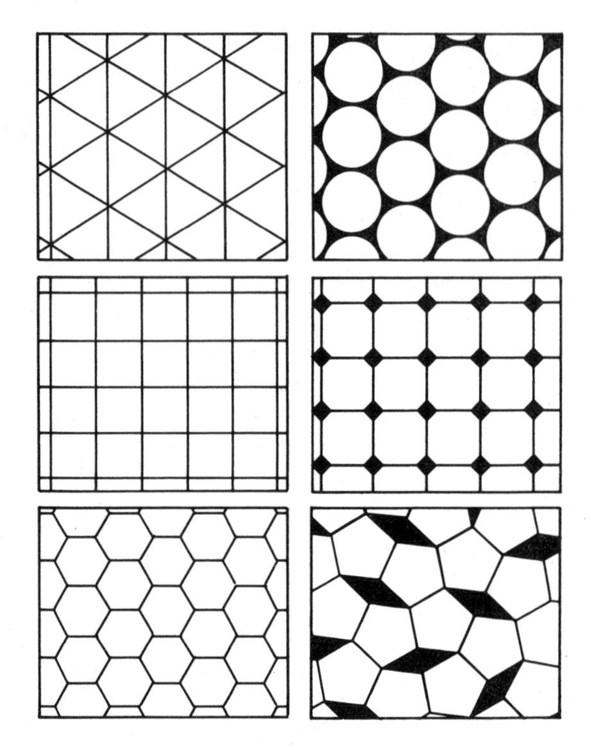

Fig. 5.5

Hexagonal and other shapes for cells, showing the advantage of hexagonal cells in utilizing area per unit volume. (Redrawn from von Frisch, 1974, in Karl von Frisch and Otto von Frisch, *Animal Architecture.* Drawings copyright © 1974 by Turid Holldobler. Reproduced by permission of Harcourt Brace Jovanovich, Inc.)

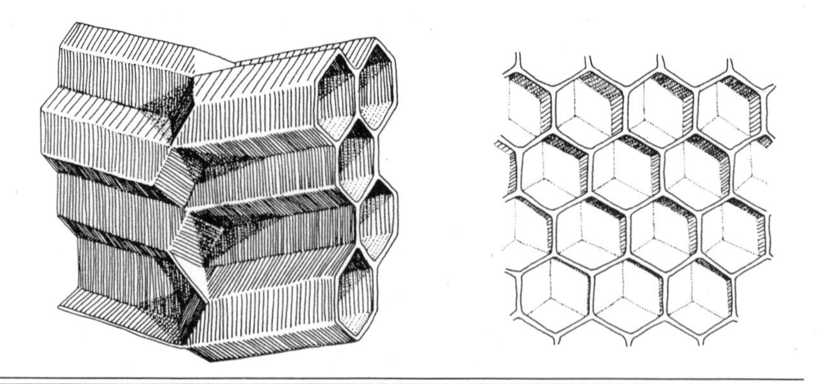

Fig. 5.6

Detail of comb, showing back-to-back comb construction and angled cells. (Redrawn from von Frisch, 1974, in Karl von Frisch and Otto von Frisch, *Animal Architecture.* Drawings copyright © 1974 by Turid Holldobler. Reproduced by permission of Harcourt Brace Jovanovich, Inc.)

that swarms of tropical-evolved bee races build smaller cells initially and wider cells when colonies mature. Studies in Peru showed that the first cells constructed by feral Africanized colonies were 4.6–4.7 mm in diameter, and cells constructed just a few weeks after colonization were 4.8–4.9 mm (Taylor, Winston, and Otis, unpublished observations). This phenomenon has never been seen with temperate-evolved bees, and it may be an adaptation unique to tropical honey bees which allows them to rear smaller workers early in colony cycles, thus conserving resources until colonies can gather sufficient pollen and nectar to produce larger cells and larger workers.

The process of comb construction has been well summarized by Ribbands (1953), von Frisch (1974), Michener (1974), and Gary (1975). When construction begins, the workers hang together in tight chains, forming a dense cluster in which they maintain a temperature of 35°C, the best temperature for wax secretion and manipulation (Darchen, 1962). Flakes of wax are removed by the enlarged first tarsal joint of the hind leg from four paired glands on the underside of the abdomen and passed forward for construction manipulations by the front legs and mandibles (Fig. 5.7). The wax is mixed with saliva and kneaded to the proper consistency and degree of plasticity at which it can best be molded. The process of removing and manipulating each scale takes about 4 min, or 66,000 bee hr to produce the 77,000 cells which can be constructed with 1 kg of wax.

Construction progresses in a seemingly random fashion, since several bees contribute to the building of any one cell, and several cells are under construction simultaneously. Workers begin construction on the roof or side of the nest cavity, with perhaps two or three construction sites initially for each comb. Thick layers of wax are first placed at the base of what will be each comb, and these are gradually drawn out into cells by elongating and thinning the wax into the cell walls. A single worker may add wax or smooth it, moving from cell to cell and building site to building site in no evident order. Apparently, each worker can perceive the stage of construction at each new location and contributes whatever is needed for that cell. As a result of this series of individual actions, each section of comb is linked with the others so that no traces of their separate beginnings are visible. The precision and strength of the newly built comb is remarkable. For example, cell wall thickness is 0.073 ± 0.002 mm, the angle between adjacent cell walls is an exact 120°, and each comb is generally constructed 0.95 cm from its neighbor (von Frisch, 1974). Furthermore, 1 kg of wax can support 22 kg of honey, over 20 times its own weight (Ribbands, 1953).

Workers do have specialized "tools" which permit such fine cell construction (Martin and Lindauer, 1966; von Frisch, 1974). The hair plates at the base of the neck serve as a plum bob to determine the

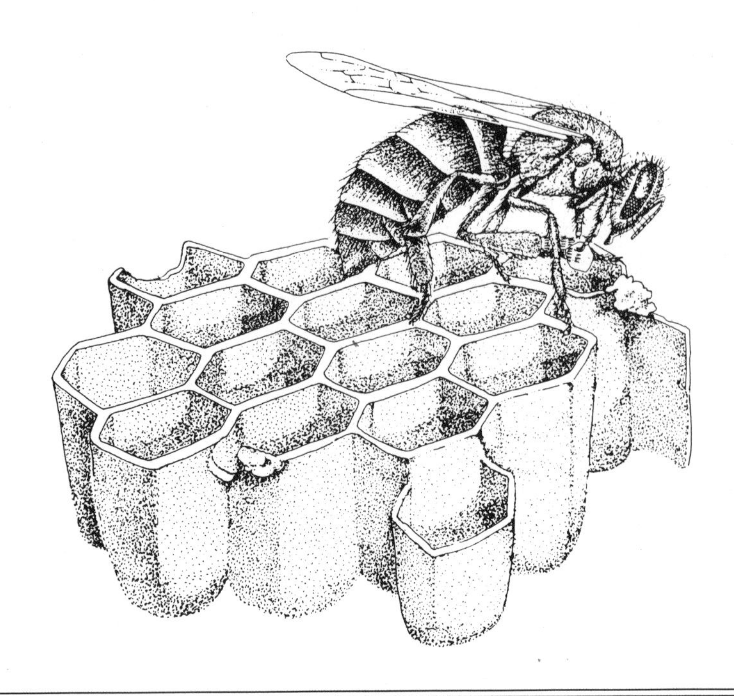

Fig. 5.7

Wax manipulation and cell construction by a worker.

line of gravity. When a worker turns relative to the earth's gravitational field, the shifting pressure of the hair plates on the associated tactile organ tells the worker which way is up. Without these hair plates workers cannot even begin to build comb (Gontarski, 1949). Interestingly, a small group of workers transported into space in the 1984 Challenger space shuttle flight was able to construct usable cells while in a weightless condition, although, given our experiments here on earth, they should not have been able to do so (Vandenberg et al., 1985).

Workers also use sense organs at the tips of their antennae to determine the thickness and smoothness of the cell walls, but the structures which control the diameter and orientation of the walls are not known. They do not appear to involve the antennae, since workers with both antennae removed construct cells with the proper diameter and angle between adjacent cell walls. Experiments with queens suggest that the front legs may be involved in cell diameter measurements. The queen must be able to sense the different diameter of worker and drone cells, since she generally lays fertilized eggs in the worker cells and unfertilized eggs in the drone cells. When the tips of the queen's legs are amputated, she continues to lay eggs but can no longer differentiate between cell sizes and produces a mixture of fertilized and unfertilized eggs which are not laid in the proper cells (Koeniger, 1969, 1970). This experiment suggests that workers could use the tips of their forelegs in a similar fashion to measure cell diameter, but such experiments do not succeed with workers; removing part of their legs destroys their ability to manipulate wax and do any cell construction.

Although comb is composed entirely of beeswax, bees use plant resins for other aspects of nest construction (for reviews, see Haydak, 1954, and Ghisalberti, 1979). These sticky resins, the propolis or bee glue, are collected by workers from resin-secreting plants and carried back to the nest on the pollen baskets on their hind legs. Propolis is used by workers to block holes and cracks in the nest, cement and strengthen the comb bases, coat the nest cavity with a thin insulating layer, and even "embalm" carcasses of intruders that have been killed but are too large to transport out of the nest, such as mice. It is interesting to note that the ancient Egyptians are also thought to have used propolis for embalming. Propolis may also have some antifungal and antibacterial properties, which protect the nest from infection and mold.

The organization of the completed nest depends on cavity characteristics and season, but generally honey is stored in the upper and peripheral sections of the nest, and brood is reared in the lower areas of centrally located combs (see Fig. 5.1). This arrangement places the heavier honey close to points of comb attachment, thus minimizing stress on the wax comb, while the brood is located so that temperature in the central part of the nest can be easily maintained and nursing activities concentrated in one area. Pollen is usually placed in cells next to the brood nest, where it is easily accessible to nurse bees. Queen cells, when present, are generally found on the bottom and sides of combs, or in crevices on the comb surface. Drone cells are usually found grouped on the comb edges (Free, 1967a; Taber and Owens, 1970; Seeley and Morse, 1976; Otis, 1980; Lee and Winston, 1985a), unlike other bee species which disperse drone cells throughout the worker cells (Michener, 1974). Grouping of drone cells may assist the queen in laying batches of fertilized and unfertilized eggs, thus allowing easier sex determination (Seeley and Morse, 1976). Grouping of drone and worker cells also allows comb to be more uniform and thus stronger, and may facilitate worker determination of the quantity of comb already constructed. The placement of drone comb on outer edges may be due to drone brood being more expendable than worker brood. As the outside temperature falls, the cluster surrounding the brood becomes tighter and smaller. Thus, the first areas to be left unattended are the peripherally located drone comb and brood. In addition, drones may withstand lower or less constant temperatures for development than workers, as they are larger and mature more slowly.

Once comb is built it is a permanent part of the nest, and honey bees do not tear down cells and reuse the wax as other bees do (Michener, 1974). Nevertheless, the cells can be reused, since workers fastidiously clean cells after brood emerges or when stored honey or pollen is removed, and damaged cells are often repaired (Darchen, 1968). Cell diameters may diminish slightly over a period of many years because of accumulated cocoons, cast-off larval and pupal skins, and treatment given cells to prepare for the next brood cycle. Such old comb appears brittle and is considerably darker than fresh comb (Gary, 1975), and workers reared in old cells may be smaller than those reared in new cells larger in diameter (Abdellatif, 1965; Nowakowski, 1969). The wax cappings with which workers cover pupae and honey sometimes can be recycled by thinning them prior to uncapping and then using the thinned wax flakes for construction elsewhere (Lineburg, 1923a,b; Meyer and Ulrich, 1952).

The surface of the comb is important for such worker and queen functions as clustering, pheromone deposition, and dancing. Workers use the comb as a substrate to form warm clusters around the brood, and the claws on bee tarsi are ideally suited to grasp on to the comb. The queen deposits a substance called "footprint pheromone" on the comb as she walks, a secretion that may prevent queen rearing (Lensky and Slabezki, 1981). The comb also serves as a platform for foraging dances, and the number of cell diameters traversed by a dancing worker is one of the mechanisms used to communicate distance to a food resource (von Frisch, 1967a). Also, worker dorsoventral oscillations (Fletcher, 1978b,c) and queen piping vibrations are transmitted through the comb and perceived by vibration detectors in the workers' tarsi.

Man-Made Nests

Many of the characteristics of feral nests have been incorporated into artificial bee hives for honey bee management. Prior to the mid-1850's, bees were kept in all sorts of small wooden, ceramic, or woven containers, but since the comb was fixed to the top and sides of these primitive hives, management was limited and honey could be extracted only by excising the comb (Crane, 1983). Modern beekeeping essentially began in 1851, when the Reverend L. L. Langstroth of Philadelphia designed a hive with movable frames (Fig. 5.8). Langstroth realized that the comb in feral colonies was naturally spaced at intervals of 9.53 mm (3/8 in.), and that if bees were given wooden

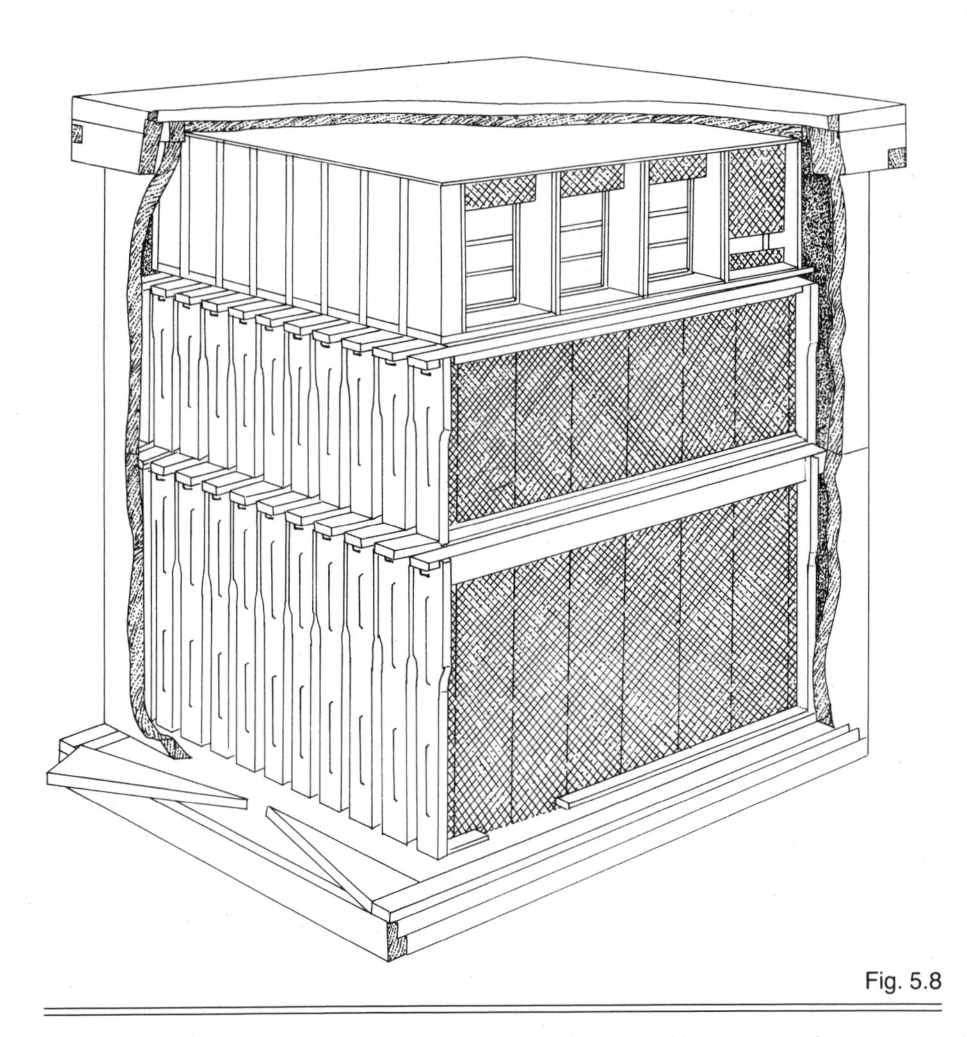

Fig. 5.8

The type of movable frame beekeeping equipment designed by Langstroth in the 1850's and still used today by most commercial and hobby beekeepers. (Redrawn from Dadant, 1975.)

frames spaced that far apart in which to build their combs, they would construct combs without brace and bridge wax between frames. Thus, beekeepers could remove the frames for examination, manipulation, and honey extraction, and then replace them in the hive. Additional boxes with frames could be stacked on top of full boxes, allowing colonies to grow to the abnormally large sizes necessary for surplus honey production. Another advantage of Langstroth hives, which has become important for modern agriculture, is that they can be moved easily without breaking combs, allowing colonies to be transported to blooming crops for pollination and to take advantage of the concentrated honey sources available as a result of intensive modern agriculture.

Also developed at about the same time as the Langstroth hive were the means of producing comb foundation and the radial honey extractor. Comb foundation consists of thin sheets of pure beeswax embossed on both sides with proper-sized hexagonal cells. These sheets can be wired into the frames, and the combs are extremely durable when drawn out and completed by the workers, often lasting for 10–20 years. Such combs can hold 3–5 kg when filled with honey, yet are capable of withstanding the centrifugal forces produced by the other important innovation for honey production, the radial extractor. When the frames are filled with honey, they can be removed from the hive and the wax cappings sliced off by passing a hot knife over the comb surface. The frames are then placed in an extractor which spins the comb at high speeds, forcing the honey out of the cells to the extractor walls, where it drops to the bottom and can then be processed and bottled. The empty combs can be reused or stored until the next season (Dadant, 1975). The processing of wax cappings has developed into an additional industry, since they are an important source of commercial beeswax, much of which is recycled by producing more foundation.

Thus, a few simple innovations derived from understanding basic aspects of honey bee nest architecture have led to the multibillion-dollar honey, wax, and pollination industries, which are the basis of modern commercial beekeeping.

6

The Age-Related Activities of Worker Bees

As their name implies, the workers perform virtually all of the tasks in the nest. From the moment of emergence, workers begin laboring at in-nest jobs such as cleaning cells, rearing brood, and building comb, and outside tasks like guarding and foraging. But there is a curious paradox which is revealed after observing colony activities; in spite of their name, workers actually spend most of their time simply standing or walking in the nest, seemingly doing nothing. These long periods of inactivity are interspersed with frantic activity sessions during which individual workers perform many different tasks in a short period before resting again. Nevertheless, a tremendous amount of work gets done in the colony, and bee society is generally perceived as orderly and efficient. These conflicting observations have inspired in-depth research in the fascinating area of age polyethism, that is, the underlying organizational factors which determine why workers do what when.

Research into the intricate nature of worker division of labor has passed through three phases. The earliest studies, conducted by Butler (1609), Donhoff (1855a,b), and Gerstung (1891–1926), determined that worker activities have a temporal basis, with in-nest tasks being performed by younger workers and outside jobs being done by older workers. These early studies led to the second phase, in which the order and nature of worker tasks were more precisely analyzed, the flexibility in temporal division of labor described, and the underlying glandular basis for the performance of many tasks discovered (Rösch, 1925, 1930; Perepelova, 1928b; King, 1933; Lindauer, 1952, 1953; Ribbands, 1952; Sekiguchi and Sakagami, 1966; for reviews, see Ribbands, 1953; Free, 1965; and Michener, 1974). The third and most recent area of research into division of labor has investigated the colony-level factors which determine why workers perform tasks at different times, and how temporal caste structure is integrated into efficient colony functioning.

The Structure of Worker Temporal Caste Ontogeny

Two major research techniques have been used to define temporal division of labor in honey bee colonies. In both types of experiments workers have been marked at birth and their activities observed throughout their lives, usually using glass-walled hives. In the first type the activities of a single worker have been monitored throughout her life by almost continuous observation, whereas in the second type many marked workers have been examined periodically to see what they are doing at different ages. These observations are extraordinarily time-consuming; Sekiguchi and Sakagami (1966) spent 720 hr observing 2700 marked bees, and Lindauer (1952) observed 1 worker alone for over 176 hr. These studies and many others have provided a wealth of data about when workers perform certain tasks, and the pattern which has emerged can perhaps be summarized as follows: Workers tend to do groups of within-colony tasks early in their lives, roughly in the order of cell cleaning, then brood and queen tending, and then receiving nectar, packing pollen, building comb, and cleaning debris from the nest. These colony duties are followed by ventilating, guard duty, and, finally, foraging trips (Table 6.1). But a few points must be noted here: first, there is considerable variability and overlap in the ages when these and other tasks are performed, and at any given age a worker will be doing a number of jobs. Second, glandular development and resorption are closely linked to many tasks, particularly brood rearing and comb building. And third, the majority of a worker's time is spent resting or walking through the colony.

The most striking characteristic of temporal caste structure in workers is the variability in the ages at which jobs are performed. For example, in various studies workers have been observed tending brood at ages of 1–52 days, with the mean age of this task ranging from 7 to 13 days. The range of ages at which workers forage was even more striking, from 3 to 65 days, with the mean ages for this task ranging from 18 to 38 (see Table 6.1). Some of this variability is undoubtedly due to differences in experimental techniques in the various studies. For instance, some of the researchers cited in Table 6.1 truncated their observations, in other words, did not follow the workers until the end of their lives. Furthermore, the definitions of tasks vary between studies, and some of the tasks were divided into subcategories such as the type of cell-cleaning activity or the age of brood tended. Nevertheless, the wide variability in temporal division of labor suggests that colony characteristics and requirements might influence the ages when workers perform tasks, and that timing of the ontogenetic sequence of tasks can be modified by colony needs.

Table 6.1 Ages (in days) at which tasks are performed by workers. Means are presented where available

Task	Age range	Mean age	Reference
Cell cleaning (early)	0–52	9.0	Winston and Punnett, 1982
	1–25	6.2	Seeley, 1982
	1–25	—	Sakagami, 1953
	1–26	—	Lindauer, 1952
	1–5	—	Rösch, 1925, 1927
	1–21	5.3	Perepelova, 1928b
	1–30	9.7	Smith, 1974
Capping brood	3–7	4.8	Seeley, 1982
	2–26	—	Sakagami, 1953
	1–26	6.3	Smith, 1974
	1–19	—	Kolmes, 1985a,b,c
Tending brood	1–52	12.6	Winston and Punnett, 1982
	2–26	12.8	Lindauer, 1952
	6–13	8.6	Rösch, 1925, 1927
	6–16	9.2	Perepelova, 1928c
	1–26	9.3	Smith, 1974
	1–13	6.5	Seeley, 1982
	1–26	—	Kolmes, 1985a,b,c
	2–31	—	Sakagami, 1953
Queen tending	1–49	17.1	Winston and Punnett, 1982
	1–10	5.5	Seeley, 1982
	1–52	10.7	Allen, 1960
Receiving nectar	8–14	11.2	Rösch, 1925, 1927
	10–22	14.9	Seeley, 1982
	5–28	—	Sakagami, 1953
	1–17	—	Kolmes, 1985a,b,c
Handling pollen	12–25	16.3	Seeley, 1982
	1–33	—	Sakagami, 1953
Comb building	1–52	15.2	Winston and Punnett, 1982
	2–52	15.8	Rösch, 1925, 1927
	1–17	—	Kolmes, 1985a,b,c
	0–34	—	Sakagami, 1953
Cleaning debris from hive	2–20	13.9	Perepelova, 1928b
	10–23	14.7	Rösch, 1925, 1927
	9–16	11.3	Seeley, 1982
Cell cleaning (late)			
Cell walls	1–21	13.3	Perepelova, 1928b
Smoothing edges	1–21	11.0	Perepelova, 1928b
Capping removal	4–21	18.2	Perepelova, 1928b

Table 6.1 continued

Task	Age range	Mean age	Reference
Ventilating	1–25	14.7	Seeley, 1982
	1–61	19.0	Winston and Punnett, 1982
	1–19	—	Kolmes, 1985a,b,c
Patrolling	0–60	15.5	Winston and Punnett, 1982
	1–27	10.3	Seeley, 1982
Resting	0–69	19.2	Winston and Punnett, 1982
	1–27	9.1	Seeley, 1982
Guard duty	4–60	22.1	Winston and Punnett, 1982
	10–46	—	Sekiguchi and Sakagami, 1966
	11–25	—	Butler and Free, 1952
	7–23	14.9	Moore, Breed, and Moor, 1986
First orientation flight	5–15	7.9	Rösch, 1925, 1927
	4–65	25.7	Winston and Punnett, 1982
	7–12	8.9	Seeley, 1982
First foraging trip	3–65	25.6	Winston and Punnett, 1982
	10–34	19.5	Ribbands, 1952
	10–32	20.1	Ribbands, 1952
	9–35	19.2	Ribbands, 1952
	20–41	30.2	Lindauer, 1952
	5–39	18.3	Sakagami, 1953
	10–27	20.6	Seeley, 1982
	10–59	37.9	Winston and Fergusson, 1985
	10–34	19.5	Rösch, 1925, 1927
	7–43		Sekiguchi and Sakagami, 1966

Another characteristic of temporal division of labor in honey bee workers is the performance of multiple tasks at any given age. Lindauer's (1952) bee number 107, possibly the most observed bee in history, spent parts of the eighth day of her life resting, patrolling, eating pollen, cleaning cells, tending brood, and building and capping comb. There were no differences between day and night observations. During this particular 24 hr, she spent over two-thirds of her time either resting or patrolling, and the other third working at these inside tasks. These observations of task overlap have since been confirmed by many studies in which definable caste groupings have emerged based on the ages when tasks are most commonly performed. In Figure 6.1 I have summarized the data from Table 6.1 into such groupings, which reveal four distinct but overlapping temporal castes: (1) cell cleaning and capping, (2) brood and queen tending, (3) comb building, cleaning, and food handling, and (4) outside tasks, including ventilating, guarding, and foraging. The middle two

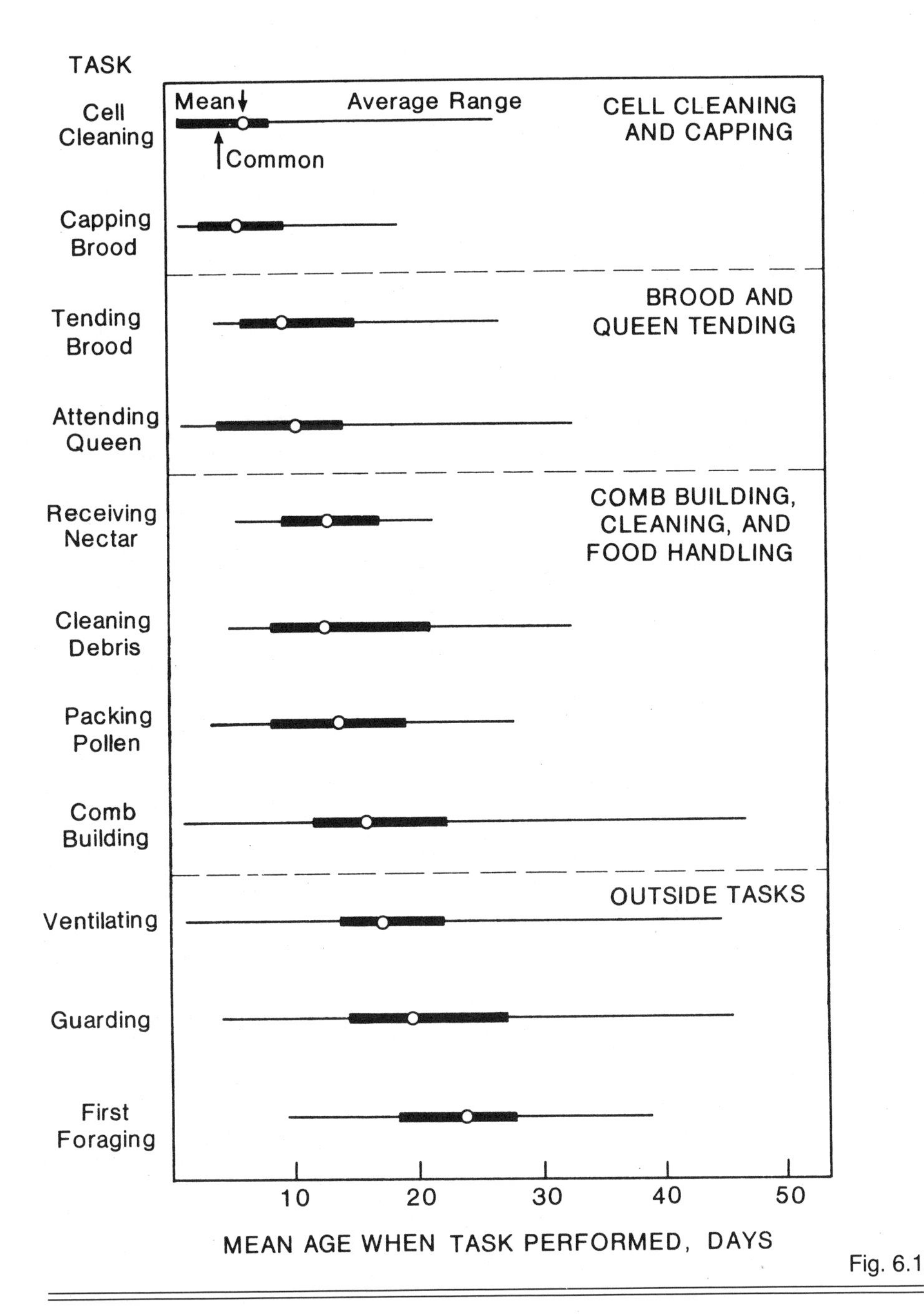

Fig. 6.1

Overlap of age-related task performance by worker bees. Data are from the references cited in Table 6.1 and represent the mean ages and most common ages when tasks were performed, and average range of ages from these studies.

groups show the most task overlap and gradual transitions between castes and may be statistically indistinguishable. Workers in the first and fourth groups tend to perform fewer tasks at those ages and tend to be quantitatively distinguishable from the middle groups (for reviews, see Kolmes, 1985a,b,c). Occasionally a worker will be observed to perform only one task during most of her lifetime, such as Robinson, Underwood, and Henderson's (1984) bee Yellow 57, which spent the last 14 days of her life collecting water, and a few workers observed by Winston and Punnett (1982), which spent much of their lives grooming other workers. However, such individual specialization is rare (Kolmes, 1985b), and the most common pattern of task performance is for workers to do a number of jobs daily except for the very young cell cleaners and the older foragers, which tend to specialize in their respective tasks.

The development and resorption of many of the worker's glands are closely linked to job performance. In a completed but never published thesis King (1933) demonstrated that the mandibular, hypopharyngeal, and wax glands all begin enlarging early in workers' lives, reaching and maintaining their maximum sizes at 5–15 days, coincidental with brood rearing and comb-building activities (Fig. 6.2). All of these glands diminish in size as their brood food and wax production decreases with the shift to food handling and other activities. In contrast, the salivary glands show a gradual increase in activity until their production peaks at about 15–25 days, after which they rapidly decline with the shift to guarding and foraging activities. During this time the hypopharyngeal glands secrete invertase for nectar processing rather than food for brood consumption (Simpson, Riedel, and Wilding, 1968). When workers begin guarding and foraging, the sting glands are producing their maximum amount of alarm chemicals, and the mandibular glands have changed from production of brood food to the alarm pheromone 2-heptanone (Shearer and Boch, 1965; Boch and Shearer, 1966).

Other physiological changes in addition to glandular development and resorption are involved in the change from nest duties to foraging (Harrison, 1986). Associated with the onset of foraging are a decrease in body mass of approximately 40%, mostly from the abdomen, and increases in the thorax-specific oxygen consumption rate and tissue glycogen levels. The effect of these physiological changes is to increase the maximum load per foraging trip and the flight range of the worker, and to decrease the energetic cost of foraging.

It is not clear whether glandular and other physiological activities are genetically programmed so that workers will be capable of performing certain tasks at certain ages, or whether colony conditions dictate gland size. But the role of colony conditions in the induction

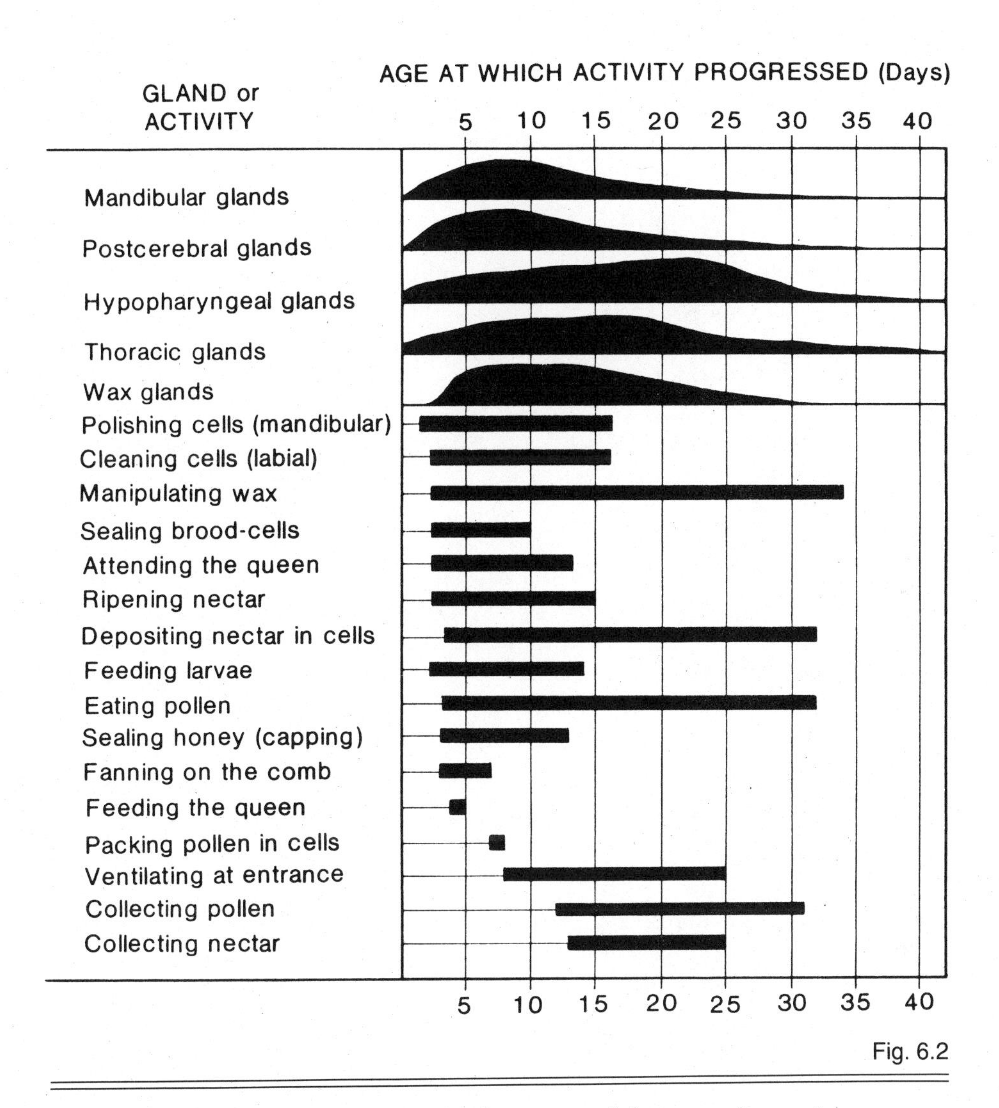

The relationship between glandular development and the ages when certain tasks are performed by workers. (Redrawn from King, 1933.)

of gland development has been demonstrated in numerous studies which have manipulated such factors as the amount of comb and brood in colonies and the numbers of old or young workers (Nolan, 1924; Rösch, 1930; Milojevic, 1940; Moskovljevic-Filipovic, 1956; Free, 1961a; Lindauer, 1961; Kolmes, 1985a,b,c). Under these manipulated conditions older workers can prolong the activity of their wax and hypopharyngeal glands or regenerate them and take up brood feeding and comb-building tasks again, suggesting that colony requirements may affect temporal caste activities in part by influencing glandular functioning. Thus, it appears that newly emerged bees follow a sequence of glandular development and resorption which coincides with task performance, but both the glandular condition and the performance of the related tasks can be influenced by colony requirements.

The last aspect of age polyethism which has been noted by almost all researchers is that workers spend most of their time either patrolling the nest or resting. This seemingly lazy behavior on the part of the workers is probably quite important in caste structure, however, since by patrolling, workers are presumably gathering information about colony conditions which will determine what jobs get done. As Lindauer (1952) points out, resting may also have an important function, particularly for workers, which may spend their resting time producing brood food or wax.

Thus, temporal caste structure in honey bee workers can be viewed as a flexible system in which workers follow a general progression of activities from nest to field tasks. Considerable variability and overlap of tasks are evident during this caste ontogeny, both among workers in the same colony and between colonies. Workers tend to perform groups of jobs at any given age, and task performance is closely linked to glandular condition, although workers spend most of their time patrolling the nest or resting.

Age-Related Tasks

Worker bees are both generalists and specialists. Any one job such as cell cleaning, brood tending, or foraging is highly specialized, yet a typical worker will perform many of these specialized tasks within the space of a few hours, and perhaps 15–20 such tasks during her lifetime. The following discussion of the major temporal worker activities listed in Table 6.1 is drawn from the references cited there as well as Ribbands (1953) and Free (1965).

CLEANING ACTIVITIES

Worker tasks related to nest cleaning can be divided into two groups, cell preparation and general nest sanitation. Cell preparation is the first activity that workers participate in upon emergence, often beginning to clean the bottoms of cells in the area from which they have emerged when only a few hours old. This early cell cleaning activity involves removing the remains of cocoons and larval excreta, and covering any remaining material with a thin layer of wax. Somewhat older workers participate in other cell-cleaning activities, such as cleaning cell walls, smoothing the cell edges, and removing cappings. A typical cell takes a bee about 41 min to prepare, with 15–30 workers taking part before the cell has been fully cleaned. The rapid completion of cell preparation is important for the colony, since the queen will only lay eggs in cells which have been cleaned. The cell-cleaning activity of young workers is responsible for the regular pattern of brood in colonies; the newly emerged workers clean cells in the area from which they emerged, and, as a result, the queen lays eggs in that part of the nest. This synchronizes the emergence of the next batch of brood from that comb, and the cycle then starts again.

Workers also remove debris from the nest, including moldy pollen, old cell cappings, and dead brood or adults. These hygienic tasks are generally performed during the same period as cell wall cleaning and smoothing, at about 11–15 days of age. Some of these cleaning bees specialize for a few days in removing dead bees from the nest, flying a short distance with the corpses before dropping them. Approximately 1% of workers in a colony conduct these undertaking duties at any one time. They appear to recognize dead bees using chemical cues that are released soon after the death of a bee (Visscher, 1983).

BROOD TENDING

The major activity associated with brood tending is nursing, in which workers with well-developed hypopharyngeal and mandibular glands feed brood food to larvae. There are conflicting reports on the age at which workers start feeding larvae, probably because it is difficult to distinguish between an inspection of a larval cell and actual feeding. Still, most studies indicate that the brood food glands are generally well developed by the time workers are 3 days old, and brood tending chores seem to peak when the bee is between 6 and 16 days old. The earliest studies suggested that younger workers tend to feed older larvae and vice versa (Lineburg, 1924; Rösch, 1925; Perepelova, 1928c), but later research has shown no differences in the ages of bees visiting larvae of all ages. A single larva is tended by many nurse bees, and larvae are visited and inspected much more frequently than they are fed. Lindauer (1952) observed that an average larva was inspected 1926 times for a total of 72 min, but only fed during 143 visits. The time per feeding visit averaged 1.3 min for a total of 110 min feeding time per larva, or slightly under 2% of its larval life. Other studies (Lineburg, 1924; Kuwabara, 1947) have shown higher values of up to 7200 visits or a maximum of 1140 feeding visits per larva, possibly because the amount of inspection and feeding may vary according to the ratio of larvae to nurse bees. When the ratios of brood to nurse bees have been calculated, results show that a single nurse bee rears the equivalent of two to three larvae during its nursing life.

The mechanism by which nurse bees determine how much food a larva requires is not known, but presumably is determined during inspections, when the nurse bee has her head inside the larval cell. All feedings are preceded by such a 2-to-20–second inspection. Larvae are fed not by direct mouth-to-mouth food exchange, but by the nurse bee placing a droplet of the brood food secretion on the cell wall or bottom, usually near the larva's mouth. The proportions of brood food from the hypopharyngeal and mandibular glands fed to larvae of different ages consistently differ, which suggests that nurse bees are able to determine larval age and food requirements and feed accordingly.

QUEEN TENDING

Workers attend the queen at about the same time they are participating in brood-nursing activities. A circle of six to ten attendant workers is usually formed around the queen, with the individual attendants rotating frequently; a typical visit lasts less than 1 min (Allen, 1957, 1960). These workers can be seen examining the queen with their antennae and forelegs, licking her with their tongues, and feeding her brood food secretions by direct mouth-to-mouth food exchange. Workers spend less time feeding the queen in winter than in summer, when she is actively producing and laying eggs. During peak egg-laying periods she is fed every 20–30 min, for 2–3 min per feeding, and generally by workers younger than 12 days old (Allen, 1955). The intense antennation, licking, and foreleg contacts made by the workers to the queen's body have two functions, grooming and pheromone perception and transmission. The latter function is carried out by workers as they move through the colony after attending the queen and antennate other workers, presumably transmitting some of the queen's pheromones.

COMB BUILDING

The construction period of a worker's life usually has two stages, cell capping by young workers and comb construction by older workers. The early capping behavior is possible since workers as young as 2–3 days old can produce wax, although the glands are most highly developed in workers 8–17 days old. Also, older wax-producing workers place wax scales on the rims of cells which need to be capped, so that copious wax production by the younger cappers is not necessary, and workers can shape and manipulate wax whether or not they are secreting wax themselves. Capping is a somewhat unorganized process in which many workers each do a small bit of capping construction in an unsystematic fashion. A typical cell might take over 6 hr to cap and have hundreds of workers participating in capping construction. Not all workers help to complete the task; often a worker will remove a piece of wax from one partially capped cell and add it to an adjoining cell capping. Cell capping can proceed more quickly, however; Smith (1959) noted a capping time of only 20 min for one cell by a single worker.

The actual construction of comb (described in Chapter 5) is performed by older workers, at ages averaging only a few days older than those for brood-tending bees. A typical comb builder might spend some of her time in a comb-building cluster, then move to the brood area and inspect or feed larvae, and perhaps do other cleaning or food-handling chores before returning to comb building. This alternation of tasks provides time for the wax glands to produce more wax for construction, and for the hypopharyngeal and mandibular glands to produce brood food.

FOOD HANDLING

Workers receive nectar and pack pollen in cells at about the same age they build comb, peaking in these food-handling activities when 11–16 days old. A receiving worker accepts nectar from returning foragers by sipping the regurgitated liquid with her tongue from the mouthparts of the donating foraging bee. The exchange takes only seconds, and a donating forager usually splits her load among two or three receivers. Once a worker has received nectar, she usually moves to an uncrowded part of the nest and repeatedly folds and unfolds her mouthparts, exposing the nectar to the air and evaporating some of the water from it. After about 20 min of this, she deposits the partially evaporated nectar into a cell, where the drying is continued by fanning bees until the nectar contains less than 18% water. This ripening of nectar takes anywhere from 1 to 5 days, depending on the water content of the nectar, humidity, nest ventilation, the amount of nectar being handled, and the number of workers involved (Park, 1925b, 1927, 1928a,b).

Nest workers also handle pollen, but after it has been deposited in cells by returning foragers. The loose pellets are moistened with regurgitated honey and saliva and packed into the bottom of cells by pushing the pellets with the worker's mandibles (Parker, 1926). The packed and moistened pollen is often covered by a thin layer of honey, and pollen stored in this fashion can be preserved for many months until needed.

VENTILATION

The first outside task that many workers perform is ventilation: they stand at the nest opening, facing the entrance, with their abdomens pointed down, and fan. Although workers of any age may be found fanning, this activity peaks when the bee is about 18 days of age. Nest ventilation may be performed for a number of purposes, including cooling the colony, evaporating honey, and decreasing the humidity and carbon dioxide levels inside the colony. This activity is most noticeable on warm summer afternoons when large amounts of nectar have been collected. At that time hundreds of workers may be found fanning at the entrance, spaced just far enough apart so that their vigorously moving wings do not touch their neighbors', and creating an audible hum which delights beekeepers. When necessary, workers also fan within the nest, particularly on comb while evaporating water from nectar.

GUARD DUTY

Guarding the nest entrance is a transient activity which some but not all workers participate in, usually before initiating foraging but sometimes afterward. Those who do guard the nest perform this activity most frequently between the ages of 12 and 25 days, and they usually guard for only a few hours or days before foraging commences (Moore, Breed, and Moor, 1986). Guard workers can be recognized by their characteristic posture at the entrance, where they stand on their four hind legs with antennae held forward and forelegs lifted. Each guard patrols a limited area around the entrance, inspecting incoming workers with the antennae and determining by their odor and behavior whether they are colony members or not (Kalmus and Ribbands, 1952; Breed, 1983b). Nestmates are readily admitted, as are young or submissive workers or dominant foragers from other colonies carrying nectar or pollen loads. More workers assume guarding duties when the colony has been under attack or during periods of forage dearth, when robbing is more likely (Butler and Free, 1952; Ribbands, 1954).

ORIENTATION FLIGHTS

The most variable age-related activities workers perform involve flight from the nest, including orientation flights and foraging. The mean age at which orientation flights are performed is usually about 1 day before the mean foraging age, but this simple calculation obscures the wide range in ages when flights occur (see Table 6.1). The purpose of orientation flights is, as the name implies, to orient to the nest location before workers take trips further afield to forage. These flights tend to take place on warm, windless, sunny afternoons, and the synchronous emergence of many young workers from the nest inspired German researchers to call these "play flights." They do have a purpose, however, since the gradually increasing circles in the air which are performed during orientation flights serve to familiarize workers with landmarks around the nest and with the location of the entrance. A single orientation flight generally lasts less than 5 min, and successive flights appear to increase in duration and distance from the colony. Remarkably, the stem length of certain brain interneurons shorten during the first flight, perhaps preparing the worker's nervous system to record and remember orientation stimuli (Coss and Brandon, 1982). Workers will void their feces during one of their first flights. (One of the hazards of sitting at nest entrances making orientation flight observations is the rapid accumulation of orange spots all over the white bee suits favored by honey bee researchers.)

FORAGING

The final task performed by workers before their death is foraging, although workers occasionally revert to other tasks, as colony needs dictate. Again, the ages of initial foraging flights are highly variable (Table 6.1), but at about day 23 of the bee's life, this activity peaks. Foragers leave the colony to collect four resources: nectar, pollen, water, and propolis. Colony conditions as well as what resource scouts and foragers encounter determine which one or combination of these a forager will collect. The brood food and wax glands of foragers degenerate, and workers who have been foraging for more than a few days begin to look old and worn-out, as they lose their hair and show wing fraying. The life span of a foraging bee is short; an average worker forages for only 4 or 5 days before she dies. Most foragers make about 10 trips a day, but Ribbands (1949) observed one worker collect 29 pollen loads in 1 day. One-hundred and fifty trips per day to artificial syrup dishes (Butler, Jeffree, and Kalmus, 1943) and 110 trips to collect water (Park, 1928b) have also been recorded. The flight distance accumulated by a forager has more of an influence on her life span than chronological age, since workers seem to die after flying a total of about 800 km, whether that distance was flown in 5 days or 30. This appears to be caused by a breakdown in the enzymatic mechanisms which metabolize carbohydrates into glycogen. When the glycogen reserves which accumulate in the flight muscles of young workers are exhausted, the older foragers are unable to synthesize additional glycogen, and they die (Neukirch, 1982).

Factors Determining Temporal Division of Labor

It is clear from the discussion above that there is some pattern to temporal division of labor in honey bee colonies, albeit a variable one. This description of age-related tasks is not completely satisfying, however; the high level of coordination apparent in honey bee colonies suggests a more profound level of organization than simply gradual transitions between highly overlapping task groupings. The search for the underlying organizational factors that control age polyethism has, in fact, revealed an elegance of function and coordination in worker caste structure which has provided considerable meaning to our observations of age-related worker task performance. It is now apparent that workers adjust their temporal division of labor schedules in response to a broad range of nest and external environmental conditions, and these adjustments provide a flexible caste system exquisitely tuned to colony requirements. Both genetic and environmental components are involved, and such factors as worker population, the amounts and type of brood and stored resources, comb-building needs, spatial organization of the colony, weather, available forage, characteristics of worker physiology, and even the colony's racial origin all interact to determine what a worker will do at any time.

The genetic component to caste determination was demonstrated in experiments with two bee races that had been introduced to South America, the tropical-evolved Africanized bee *A. m. scutellata* and the temperate-evolved European bee *A. m. ligustica* (Winston and Katz, 1981, 1982). To separate the heritable and environmental components we used the technique of cross-fostering, in which newly emerged workers of both races were marked and introduced into colonies of both races, and the ages the introduced workers initiated foraging were compared. The results were striking; in colonies of their own race, the tropical Africanized bees began to forage significantly earlier than European bees did in European colonies, 20 versus 26 days. Even more impressive were the results from cross-fostered workers. Africanized bees in European colonies began foraging at older ages than they did in their own colonies, 23 days old on average, and were not statistically distinguishable under those conditions from European workers in those colonies. The opposite situation, in which European workers were compared to Africanized workers in Africanized colonies, revealed that the European workers began foraging even earlier than the Africanized bees, at 14 days of age.

Our hypothesis to explain these results is that the level of stimuli which induces foraging behavior is higher in Africanized colonies, and that Africanized workers have a higher threshold for those stimuli. These stimuli might include the age distribution of workers, worker longevity, colony size, amount of brood rearing, honey and pollen stores in the colony as well as availability of nectar and pollen in the field, general activity levels, and previous colony history. We made the assumption that the probability of a worker foraging increases with age, which seems reasonable in light of studies which have consistently demonstrated that older workers do most of the

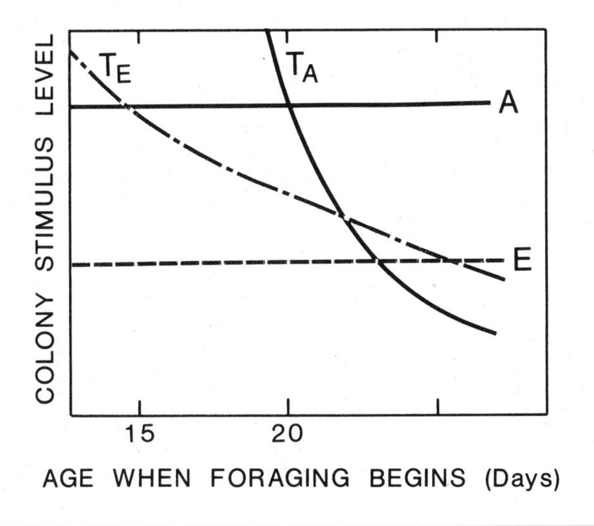

Fig. 6.3

The age at which workers begin foraging as a function of the stimulus level in colonies. Africanized colonies (A) are assumed to have higher stimulus levels than European colonies (E). T_A and T_E are the thresholds at which Africanized and European workers begin to forage. (Redrawn from Winston and Katz, 1982.)

foraging. A likely mechanism for this age-dependent shift to foraging tasks is that older workers require fewer stimuli to begin foraging than younger workers.

Differences in foraging ages between workers of the two races can be shown graphically as a function of worker age and the presumed stimulus levels in colonies (Fig. 6.3). The stimuli for foraging in Africanized colonies was seen to be greater than those in European colonies, since Africanized workers foraged at later ages in European colonies, and European workers began foraging at earlier ages in Africanized colonies. In addition, Africanized workers showed a higher threshold for foraging stimuli, in that the European workers in Africanized colonies began foraging significantly earlier than Africanized workers in the same colony environments. Thus, there are genetically based differences in the ontogeny of at least one task, foraging, between temperate- and tropical-evolved bees, and these differences involve both colony environment and worker responses to colony conditions.

One of the colony characteristics which influences temporal division of labor is colony population; this factor is particularly important in determining foraging age (Winston and Fergusson, 1985). In that study one-half to two-thirds of the workers were removed from colonies in the spring, a loss in worker population similar to naturally occurring events in feral colonies, such as predation, swarming, nest damage, and disease. Newly emerged and marked workers were introduced to those colonies and to control colonies from which no workers had been removed, and the ages when foraging started determined. Workers began foraging at younger ages and had shorter life spans in colonies from which workers had been removed; workers in control colonies initiated foraging at a mean of 38 days, while workers in colonies from which 1.8 kg and 2.7 kg of workers had been removed began foraging at 33 and 31 days, respectively. Remarkably, these and other similarly stressed colonies produced as much brood and honey as the unstressed controls by the end of the season (Winston, Mitchell, and Punnett, 1985).

These results are of particular interest because they may explain one of the functions of resting. It appears that workers resting in the colony form a reserve pool that reacts to deleterious changes in colony conditions such as swarming, pest damage, or disease, and to sudden opportunities such as discovery of an abundant nectar resource; the pool can also respond quickly to predators before nests or brood are damaged. By increasing their activity levels, workers may compress the normal ontogeny of tasks into a shorter life span; the resting workers work harder and die earlier in stressed colonies than in unstressed ones, evidence of a mechanism for temporal caste ontogeny to be adjusted in response to colony needs.

Other changes in temporal division of labor have been documented in response to colony requirements, particularly those concerned with brood rearing and construction. However, most of these demonstrated shifts in caste ontogeny have involved extreme colony manipulations, such as removing most of the adult workers and reconstituting the colony with emerging brood, or removing all of the comb. In both of these situations workers began brood food and wax secretion, nursing, and comb construction at very young ages. In colonies with only young workers orientation flights, guarding, and foraging began earlier as well (reviewed by Ribbands, 1953; Free, 1965; Michener, 1974). In situations where young workers and/or comb were removed, older workers regenerated their brood food and wax glands and actively participated in nursing and construction (Rösch, 1930; Milojevic, 1940; Lindauer, 1961). When pollen traps were put on hive entrances, preventing the entry of most pollen, workers began foraging at younger ages (Fukuda, 1960).

Less extreme manipulations can also result in temporal shifts in task performance; for example, the age when foraging commenced was inversely correlated with the amount of eggs and larvae in colonies, suggesting that workers start foraging at younger ages in colonies with more brood (Winston and Fergusson, 1985). Workers may also respond to colony conditions by increasing the time spent on a given activity without changing the ages when that task is performed. Kolmes (1985a, unpublished observations) demonstrated that in colonies from which some but not all comb was removed workers of comb-building ages spent more time doing that task but did not perform it at ages significantly different from those in nonstressed colonies. Thus, the response of temporal caste structure seems to depend on the extent of the disruption of normal colony activities. Easily documented are shifts in division of labor which can compensate for changed colony conditions under extreme stress. More subtle changes in colony attributes result in subtle changes in division of labor, sometimes involving minor shifts in the ages when tasks are performed or in the amount of time spent on a task at a given age.

Another aspect of internal colony organization which influences caste ontogeny is the location of jobs within the nest. Seeley (1982) divided workers into temporally based subcastes similar to those in Figure 6.1, but suggested that the evolution of these subgroups was based in part on the locations within the nest where tasks are performed (Fig. 6.4). The youngest workers primarily clean cells because they can easily locate a long series of cells which need cleaning in the area from which they recently emerged. The next group of tasks also take place in the nest center, and include such activities as feeding and capping brood and attending the queen. As workers get older, they begin to disperse and perform activities throughout the nest, first doing tasks such as grooming and feeding nestmates and ventilating, and gradually moving to the nest periphery for food-handling chores. The final subcaste involves foraging and other activities outside the nest. Thus, the spatial organization of colonies seems to be another underlying factor which is involved in the organization of temporal division of labor.

Factors external to the colony, particularly the availability of nectar and pollen, may also affect caste ontogeny, both directly and in conjunction with colony food requirements. This concept has been most comprehensively discussed by Ribbands (1952, 1953), who suggested that the amount and quality of nectar and pollen coming into the colony could in part determine the extent of brood rearing, food handling, and comb building and the ages at which these tasks are performed. There is certainly abundant evidence that brood rearing is increased when nectar and particularly pollen are available, and that stored honey or incoming nectar is required for wax secretion and comb building. It might be expected that resource-stimulated increases in these tasks would affect the ages at which they were per-

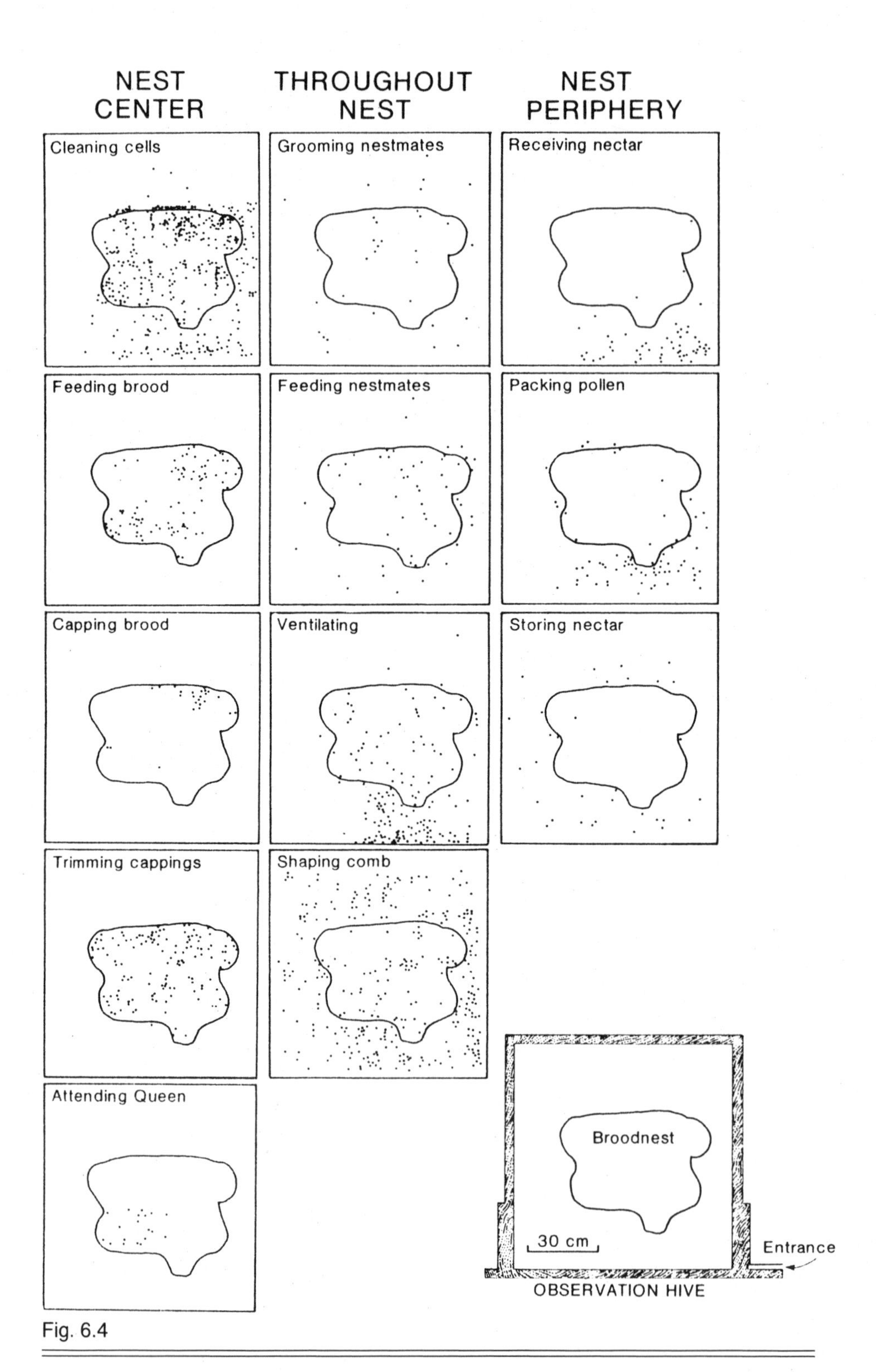

Fig. 6.4

Maps depicting work sites of some tasks within the nest, and a schematic diagram of the observation hive used for these studies. (Redrawn from Seeley, 1982.)

formed, but this question has yet to be specifically addressed. The most direct evidence supporting this hypothesis comes from Kolmes (1985a), who showed that worker bees in seasons of poor nectar and pollen availability performed fewer tasks associated with the acquisition and storage of these materials than workers during better seasons. No shifts were noted in the ages at which these tasks were performed, however. More direct studies comparing division of labor under different resource conditions are needed to examine adequately the interaction between colony environment, resources, and caste ontogeny.

The Missing Link: Perception and Control

Although there are still many aspects of these task patterns which need to be examined, it is clear that the flexible and overlapping nature of caste ontogeny is responsive to at least some colony conditions and provides a mechanism for colonies rapidly to reallocate their work performance in light of changing colony requirements. There is another aspect of caste ontogeny which needs to be more fully examined, however, to complete our understanding of age polyethism, and that is the perceptual and control mechanisms by which workers become aware of and act to satisfy colony needs. There are four stages in controlling task performance, by which (1) colony requirements are (2) perceived by workers, and (3) those perceptions are translated by physiological control mechanisms into (4) the ages when tasks are performed (Fig. 6.5). Little is known about the middle two steps, the perceptive and internal worker control mechanisms which mediate age polyethism.

The nature of the perceptions by which colony needs are translated into tasks has never been explicitly examined in the context of age-specific work performance. But studies in other contexts suggest that perceptual mechanisms such as patrolling, pheromones, trophallaxis, rate of incoming resources, and dances are involved in the control of tasks. Workers purposefully patrol the nest and appear to be assessing the colony state, although the precise age-related cues they are using have not been determined. The level of certain pheromones in the colony probably influences worker activities; for example, an as yet undescribed brood pheromone stimulates foraging for pollen (Free, 1967b) and might influence the ages when workers begin that task. There is considerable trophallaxis between workers, which could provide information on food-handling requirements and the rate of incoming resources. The waggle and dorsoventral abdominal vibrating dances, among others, provide considerable information about forage quality and quantity which might influence age-specific

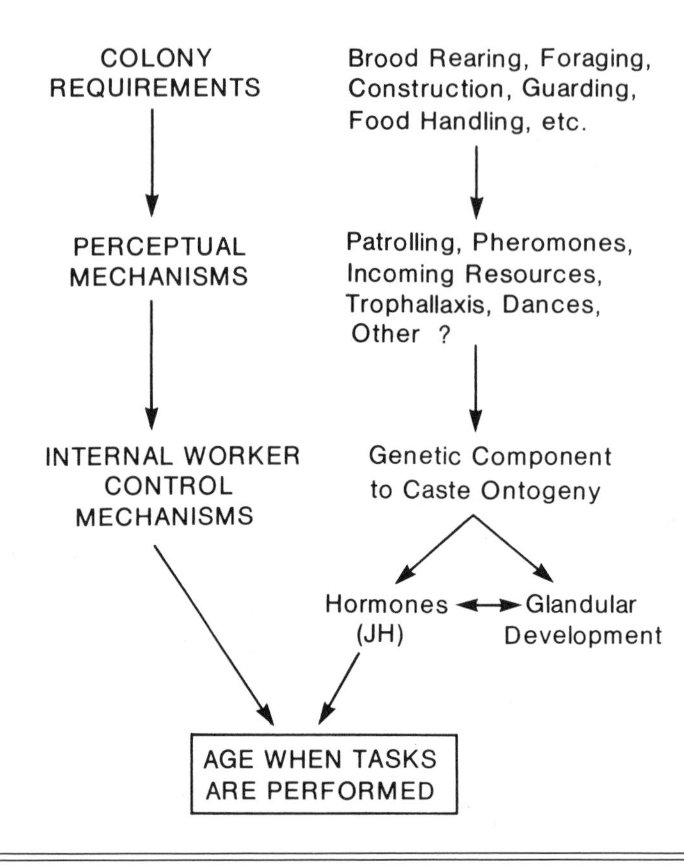

Fig. 6.5

The factors which control the ages when workers perform tasks. Arrows indicate direction of flow from colony requirements through controlling factors that determine the age of task performance.

tasks, and other types of communication, such as sound production or grooming, could transmit information used in determining temporal division of labor.

Of these proposed perceptual mechanisms, trophallaxis has received the most attention in the context of age polyethism. The extent of mouth-to-mouth food exchange was first demonstrated by Park (1923b), Rösch (1925), and Nixon and Ribbands (1952), who showed that over half of a colony's workers had colored or radioactively labeled nectar in their honey stomachs only 24 hr after less than ten foragers had brought it into the colony. Foragers were particularly involved in food exchange, and over 75% contained radioactive nectar within 24 hr. Free (1957) and Korst and Velthuis (1982) further demonstrated five key points. First, there is a tendency for food to pass from older to younger bees. Second, exchange occurs between bees with almost equal amounts of food in their crops. Third, the previous availability of food influences participation in food exchange. Fourth, some workers tend to beg for food, while others specialize in offering it. And fifth, many trophallactic interactions result in little or no actual food transfer. All of these authors have speculated that trophallaxis is involved in age polyethism, although the exact mechanism has yet to be elucidated. Nevertheless, the characteristics of food exchange described here suggest that during trophallaxis some information is transferred which could influence temporal aspects of division of labor.

Whatever the perceptual mechanisms involved in mediating task performance, internal worker control mechanisms link those perceptions to temporal division of labor. These internal factors include a genetic component which interacts with hormonal secretion and glandular development (see Fig. 6.5). We have already seen that caste ontogeny has an underlying genetic basis, and that the development and resorption of brood food, wax, and pheromone-producing glands is closely linked to division of labor. It is becoming increasingly apparent that hormones, particularly juvenile hormone, are also involved in age-specific task performance. Workers treated with a JH mimic or analogue compound shifted from the broodnest to the food storage region of colonies earlier than untreated workers (Jaycox, Skowronek, and Gwynn, 1974; Jaycox, 1976; Robinson, 1985). The amount of JH also was related to hypopharyngeal gland development; removal of the JH-producing corpora allata stopped hypopharyngeal gland development (Imboden and Luscher, 1975), and applications of JH or its analogues induced premature degeneration of those glands (Rutz et al., 1976, 1977). Finally, Robinson (1985) demonstrated that application of a JH analogue induced workers to display early orientation and foraging behavior and to produce alarm pheromones earlier than untreated workers. All of these results support the hypothesis that high titers of JH are involved in the degeneration of brood food glands and the shift from nest to field duties. Future research to unravel the perceptual and hormonal cues involved in determining temporal division of labor will certainly be a high priority during the next few years, particularly as this type of research expands to encompass the interaction between colony characteristics and age-dependent tasks.

7

Other Worker Activities

It is clear not only that workers are able to perform a remarkable number of age-related tasks, but also that they are flexible in the ages at which they do many jobs and thus can adjust their work schedules to colony requirements. But many other activities are required for normal colony functioning, some that must be performed on a regular basis, and others that are performed in response to rapid changes in colony needs. In a typical colony day workers might spend some of their time warming or cooling the colony, defending the entrance from a skunk attack, and then discovering that the colony in a nearby log is weak and flying over to rob its honey. Some of these workers might not orient properly in returning to their own nest, perhaps drifting into a neighboring colony which has just lost its queen. Workers respond to all these and other circumstances through slight behavioral modifications adapted to each unique situation.

Nest Defense

We tend to think of bee colonies as being "aggressive" or "gentle," but in fact they are neither. Honey bees have evolved numerous mechanisms for colony defense, and the level of aggression manifested by a colony is simply a measure of the sensitivity and intensity of defensive behavior. Honey bees have good reason for defending their colonies, since nests provide a rich and concentrated source of food, including brood, wax, honey, and pollen, in addition to the adult bees themselves. The predators and parasites which attack bee colonies are many and varied, and the level of defense is undoubtedly related to predation pressure in regions where different bee races evolved. The tropical African habitats where honey bees first evolved have the most predators, and it is not surprising that African races exhibit much higher levels of defensive behavior than do more temperate-evolved races.

Vertebrates, particularly humans, are some of the most serious bee predators. Rock paintings many thousands of years old have been found which depict primitive honey-hunting techniques still used today (Crane, 1983). Bees nesting in logs or out in the open are typically smoked first with a torch to pacify and confuse the workers, and then the honey hunter reaches into the nest and removes comb containing honey, often receiving many stings in the process. Honey hunting has been so widespread and common in Africa and Europe that it undoubtedly has been one of the selective pressures favoring quick and intense defensive responses by workers.

Other vertebrate predators which attack bee nests and bees in flight include bears, honey badgers, skunks, and birds. Bears can be a serious problem for bees, since their powerful legs and claws can easily tear open most wild nests, and their thick fur makes them almost oblivious to stings. Once a bear has discovered the tasty brood and honey in a nest, it will continue to attack other nests whenever they are encountered. The honey badger is an important tropical predator, and its tough skin and strong legs allow it to tear open nests without being dissuaded by worker attacks (Walker et al., 1975). The badger has an interesting relationship with the aptly named "greater honey guide" bird, which leads badgers, and sometimes humans, to nests and indicates the exact nest location by a series of churring sounds. Once the nest has been opened, the honey guide consumes the wax, which it is capable of digesting and utilizing for energy (Queeny, 1952; Friedmann, 1955); the brood and honey are eaten by the coattackers. Skunks scratch at the entrance of nests at night and eat the adult bees which respond to the disturbance; a successful skunk may return to the same colony night after night, seriously diminishing the adult bee population (Storer and Vansell, 1935; Eckert and Shaw, 1960). Adult bees flying in the vicinity of the nest entrance or out foraging or mating are eaten by many birds, including bee eaters, woodpeckers, titmice, shrikes, swifts, and tyrant-flycatchers (Ambrose, 1978). Other vertebrates that prey on bees are the giant toad *Bufo bufo,* which eats bees at the nest entrance, opossums, shrews, hedgehogs, armadillos, and anteaters (Caron, 1978; Morse, 1978).

Many invertebrates also attack bee colonies, again with more intense predation pressure in tropical habitats. The major groups of insects which attack nests, the ants and wasps, are in the same order as bees; army ants and some of the social vespine wasps are particularly important. Army ants attack the nest in numbers which rival the bee colony population and, when successful, carry off all the brood, leaving behind piles of dead adults. Wasps hover at the entrance or near flowers, attacking adults and removing the thorax and carrying it back to the nest for feeding the wasp larvae. Sometimes the wasps gain entry to bee nests and consume brood; in 1949 the secretary of the Israel Beekeepers Association reported a loss of 3000 colonies due to attack by vespid wasps (De Jong, 1978). Other honey bee colonies can also be serious predators, particularly in robbing honey. Moths attack bee colonies as well, consuming stored pollen, honey, and even wax; the greater and lesser wax moths are capable of digesting beeswax and are major pests in nests throughout most of the world (Williams, 1978). In short, an undefended honey bee nest would have almost no chance of surviving vertebrate and invertebrate predation, and stinging and other defensive behaviors are essential for colony survival.

The defensive system begins when predators are recognized at the entrance by guard bees or other workers who happen to be there, which alert their nestmates by their tense posture and release of alarm chemicals from the sting and mandibular glands (Ghent and Gary, 1962; Maschwitz, 1964a). The alerted nestmates then leave the colony and search for the attacker, orienting to such stimuli as motion, color contrasts, vibration, and scent. Once a predator is found, a worker often performs a number of anti-predator behaviors prior to stinging, since she loses her life once she stings. These prestinging behaviors can include threat postures, buzzing, burrowing into the predator, biting, pulling hairs, and running (Collins et al., 1980). Once a perceived attacker is stung, other workers orient quickly to the sting site, using alarm pheromones given off by the sting glands. Dark colors, rough textures, animal scents, and rapidly moving objects all induce workers to sting, and they are characteristics of many of the potential predators of bee colonies (Free, 1961b).

The intensity and nature of colony defense can be attributed to a genetic component and such environmental factors as temperature, humidity, colony size, worker age distribution in the colony, and availability of nectar, pollen, and water in the field. The genetics of defensive behavior has been the most well-studied factor, since racial variability in this trait is extreme. To evaluate a colony's defensive behavior, tests have been devised in which a standard stimulus is presented to a hive, usually by swinging a leather ball at the entrance following a disturbance such as a marble shot at a hive or presentation of alarm pheromones on leather balls. Factors such as the time to various arousal stages, the number of stings in the leather ball, and the time until the hive quiets are measured (Stort, 1974a; Collins and Kubasek, 1982). From such studies it is clear that tropical bee races have much higher levels of defensive behavior than temperate races, many defensive traits being up to ten times as intense in the tropical bees. In one study Africanized bees in South America reacted to a standard stimulus in 1–5 sec, whereas European-evolved bees took 10 sec or longer. And the average number of stings in a leather ball was about 80 from Africanized colonies and only 10 from European colonies (Collins et al., 1982). In another series of experiments the leather ball received an average of 61 stings and the experimenter's gloves 49 stings in 1 min at the entrance to Africanized bee colonies; values for European colonies were 26 and 0 stings, respectively, in the same time period (Stort 1974a,b, 1975a,b,). Such quantitative observations are supported by qualitative experiences of beekeepers and others in contact with bees, who report ferocious attacks by tropical bees following only minor disturbances and many deaths of livestock and humans due to the rapid colony mobilization and large numbers of stings received. In contrast, temperate-evolved bees are slow to arouse under most circumstances, and an intruder will receive few stings following colony disturbance.

The genetic basis of defensive behavior has been demonstrated within single bee races through hybridization of aggressive and gentle lines. Boch and Rothenbuhler (1974), for example, crossed aggressive and gentle lines and determined that the resultant hybrids were more gentle in some characteristics, indicating dominance of the gentle traits, and intermediate in others, indicating lack of dominance in either direction. But subsequent experiments by Collins (1979) and Stort (1974a,b, 1975a,b) suggest that aggressive behaviors may in fact be dominant, since their first generation hybrids retained much of the intensive defensive behaviors of the more aggressive lines. It is clear from their studies that a number of genes are involved in nest defense, with two or three loci controlling each of the components of defensive behavior. Again, beekeepers' observations of wide variability in aggressive behavior from colonies headed by related queens kept under identical conditions support the existence of a strong genetic component to colony defense.

Environmental factors also have a role in determining the intensity of defensive behaviors. Higher temperatures are associated with increased probability of a defensive response and with higher speed, intensity, and duration of responses; higher humidity seems to affect only the response intensity (Collins, 1981; see also Drum and Rothenbuhler, 1984). Colonies with more empty comb also show higher levels of defensive behavior, possibly because primer pheromones for colony defense are produced by comb (Collins and Rinderer, 1985).

Other factors which may influence defense are colony population, amount of stored resources, availability of nectar and pollen in the field, and previous experience. However, data concerning these factors are either unavailable or ambiguous. It is considered a universal truth by beekeepers that large colonies are more aggressive than small ones, which makes sense since large colonies have more to defend and more workers to expend in nest defense. In one study (Collins et al., 1982) the results for both European and Africanized bees confirmed this prediction: larger colonies reacted faster and with more stinging behavior than smaller colonies. Other studies have shown no positive correlation, and occasionally a negative relationship, between colony size and defensive behavior (Boch and Rothenbuhler, 1974; Collins and Kubasek, 1982). Beekeepers' observations suggest that colonies with little stored honey are less aggressive than those with more honey to defend, and that all colonies are less aggressive when there is good honeyflow in the field, possibly because many of the older bees are out foraging. Finally, colonies which have been aroused to defense continue to be aggressive for many hours and sometimes days. All of these behaviors require additional study.

Individual workers in a single colony show considerable variability in their defensive responses, older bees generally being more sensitive to disturbances. This is partly because of their location, since older bees tend to be at the colony entrance more than younger bees, but there is also a physiological basis to responses to intruders. Older bees are more sensitive to alarm pheromones and other odors; electroantennogram recordings have shown that the nervous responses of workers age 8 days and older are about three times as sensitive as those of 1-day-old workers (Masson and Arnold, 1984; Allan, Winston, and Slessor, unpublished observations; see also Fig. 8.3). Moreover, older workers produce more alarm pheromone than young workers, reaching their maximum production at 15–30 days of age, coincidental with the onset of guarding and foraging (Boch and Shearer, 1966). Finally, juvenile hormone influences the defensive state of workers. Reduction in size of the JH-producing corpora allata is associated with reduced levels of aggressive behavior toward queens (Breed, 1983a), and application of methoprene, a JH mimic compound, induces premature production of alarm pheromones (Robinson, 1985).

Honey bees have different tactics for defense, depending on the nature of the attackers. Attempts to chase away vertebrates usually involve biting, hair-pulling, and eventually stinging. Other insects are more difficult to sting, since they have the same type of hardened cuticle as bees and can only be stung through the soft connective membranes which hold the exoskeleton together. Consequently, attacks by insects are met with more grappling behavior than those by vertebrates, and a number of workers gang up on attackers to drive them off. In response to ants, workers at the entrance turn away, fan the wings, and kick their rear legs, which frequently succeeds in preventing the ants from entering the nest (Spangler and Taber, 1970). Wasp attacks may be met by shimmering behavior, whereby workers shake violently from side to side, which again often dissuades the attackers (Butler, 1974). Some of the most elaborate defensive behaviors are actually designed to prevent other honey bee colonies from robbing (as will be discussed in the next section).

Honey bees also have various mechanisms for defense against fungi and microorganisms which can attack stored products or the bees themselves. Honey has its own chemical defense, as it contains glucose oxidase enzymes, which break down glucose and release hydrogen peroxide, an antibacterial agent (White, Subers, and Schepartz, 1963); it also has a high osmotic pressure owing to its high sugar content, which inhibits fungal growth. Stored pollen is protected in part by the thin layer of honey which covers it in cells, but some pollens may themselves contain antibiotics which provide further protection (Stanley and Linskens, 1974). Bees exhibit some behavioral resistance to bee diseases, for example, American foulbrood, which is caused by the bacteria *Bacillus larvae* and kills honey bee brood. Workers from disease-resistant lines remove dead larvae and pupae from the nest more quickly than susceptible lines, thereby reducing the source of infection (Rothenbuhler, 1964). In addition, workers from resistant lines may feed larvae an as yet unidentified antibiotic which reduces the effects of foulbrood, or may remove spores from infected honey taken into the crop by the action of the proventriculus straining out the spores (Thompson and Rothenbuhler, 1957). Worker midguts contain various detoxifying enzymes which can break down poisons encountered in the environment, such as alkaloids and other substances sometimes found in plants. These enzymes are also used to combat pesticides and other contemporary chemical poisons (Yu, Robinson, and Nation, 1984).

Robbing

Honey bees are quite opportunistic in foraging, and any rich source of honey may be quickly discovered and exploited. The richest source of honey normally available in the field is not flowers but the honey stored by another colony. Robbing behavior occurs whenever one colony's bees can gain access to another colony's nest and remove the honey. This kind of attack can last for days and leave thousands of dead bees. The behavior of guards at the entrance is largely designed to protect bee colonies from members of their own species.

Robbing generally does not occur during periods when resources are readily available in the field; at such times, guard bees often permit foreign workers to enter their nests after careful and prolonged inspection and dominance interactions. These workers are frequently adopted by the new colony. However, the behavior of potential robbers is quite different from that of drifting and disoriented foragers from other colonies, and it elicits more intense defensive behavior from guard bees. Robbers become aware of the presence of a concentrated honey source in another bee colony by perceiving the strong honey odor coming from the colony entrance. These odors are strongest at the end of the summer, when colonies are full of honey, and fanning to cool the nest and evaporate nectar is at its peak. Potential robber bees often exhibit a characteristic hovering and sideways flight pattern at the nest entrance, possibly because they (1) are not familiar with another colony as a food source, (2) perceive potential defenders at the nest entrance, or (3) are alerting other robbers to the location of the colony being robbed (Free, 1954; Gary, 1966). When this erratic flight pattern as well as the foreign odor of robbers are recognized by guard workers at the entrance, they quickly respond with ferocious attacks. The guard and robber grab hold of each other's legs, curl their abdomens, and attempt to sting each other while rolling together in a cartwheel-like motion. Usually one or the other worker is seriously injured or dies (Butler and Free, 1952; Ribbands, 1954). If the robber is successful at gaining entry to the colony, she will ingest a load of honey and return to her colony, recruiting other workers to rob the attacked colony. If the robbing colony is more populous than its victim's, it can clean out all of the honey, which can result in the death of the robbed colony. Once robbers have been trained to locate the rich honey source in other colonies, they may continue searching for additional colonies to attack. Robbers become smooth, shiny, and almost black, as a result of the occupational hazards of fighting with other bees.

Nest Homeostasis

One of the great advantages and challenges of insect social organization is colony homeostasis, or maintenance of nest temperature and other environmental factors at relatively constant levels regardless of external conditions. The advantages of homeostasis are many, including rearing of brood under stable conditions, survival of populous colonies through cold winters and hot summers, early spring initiation of brood rearing, and preflight warming of foragers. The evolution of individual worker traits that must be coordinated by colony-level organizational factors is of extreme importance, since workers must be able to respond to ambient conditions ranging from below freezing to fatally hot temperatures, often within the same day. Compounding the problem are constraints on homeostatic activities such as the physiology of individual workers, nest design, and behavioral limitations on worker activities. In spite of these limitations, honey bees are able to use a combination of nest design and worker activities to maintain superb control over the colony environment, and they have reached the pinnacle of nest homeostasis among all the social insects. The analysis below is concerned with the most important aspect of homeostasis, thermoregulation, which has been reviewed most recently by Simpson (1961), Seeley and Heinrich (1981), and Heinrich (1985).

The first mechanism for temperature control lies in the choice of nesting site by swarms, as discussed in Chapter 5. Once the cavity is colonized, bees improve their capacity for internal climate control by sealing unnecessary openings with propolis; some races which have evolved in particularly cold climates sometimes reduce the entrance in winter with "curtains" of propolis (Ruttner, 1968). Even the design of comb construction contributes to nest homeostasis, since the brood are surrounded by layers of temperature-buffering wax combs, and workers are able to expand or contract their clusters by moving together or spreading apart through the spaces between combs. In tropical climates warming the nest is often not as important as cooling it, and this may explain why many tropical nests are found in the open, suspended under shady branches or rock overhangs.

The most extreme thermoregulatory challenge faced by colonies occurs during the prolonged cold spells of winter. Honey bees survive this period by using energy derived from consumption of stored honey to generate body heat and keep the nest at an adequate temperature for adult survival. In addition, the workers begin to cluster when the ambient temperature reaches about 18°C; as outside temperatures drop, the cluster contracts (Fig. 7.1). Cluster contraction conserves heat by diminishing the surface area over which heat can be lost and by reducing internal convection currents. Below about 14°C the cluster has a compact outer shell of relatively quiet bees and an inner core where workers have more room to move about (Gates, 1914; Phillips and Demuth, 1914; Wilson and Milum, 1927; Corkins, 1930). The shell can be several layers thick, and workers are oriented with their heads inward (Farrar, 1943; Simpson, 1961). The cluster no longer contracts once the temperature has dropped to about −5°C; at this and colder temperatures workers generate additional heat rather than tightening the cluster for temperature control (Free, 1977).

Individual workers periodically consume honey to produce heat, and clusters may break occasionally to allow workers to move through the nest and feed. If conditions are too cold, colonies can die even with substantial honey reserves because workers are unable to leave the cluster to get to honey located at the nest periphery (Haydak, 1958). There is little or no brood rearing during the coldest parts

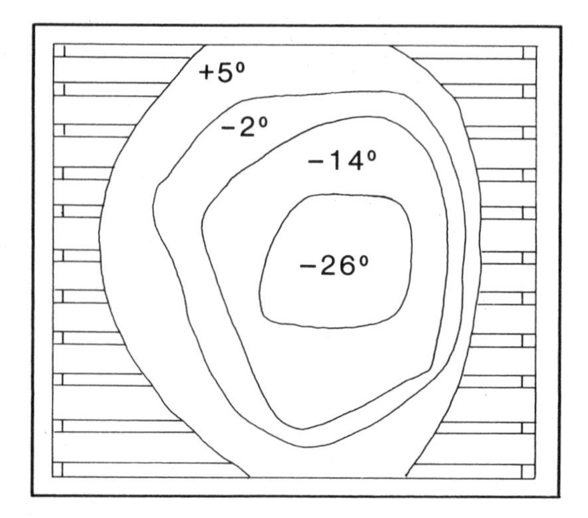

Fig. 7.1

The diameter of a cluster in a single colony under various temperature conditions (°C). (Data from Wilson and Milum, 1927.)

of winter, which means that colony temperature can be allowed to fluctuate more than it would during brood-rearing periods. Fluctuations of almost 20°C have been recorded, owing largely to changes in ambient temperatures and to workers periodically breaking from clusters to feed. The minimum temperature within the center cluster is 13°C, which maintains the temperature of the outer shell at 8°C, the minimum temperature required for workers to cling to the cluster (Hess, 1926; Wilson and Milum, 1927; Corkins, 1930; Owens, 1971; Southwick and Mugaas, 1971). Inner cluster temperatures are more commonly maintained at closer to 20°C. Although workers could undoubtedly maintain warmer cluster temperatures, the rate of honey utilization is reduced by keeping clusters close to the minimal temperatures for adult survival.

As spring approaches and brood rearing commences, the thermoregulatory requirements for colonies become much more rigid, at least in areas of the colony which contain brood. Nest temperatures in the brood area are maintained at 30–35°C, the optimal temperature range for brood rearing; temperatures in other parts of the nest are allowed to fluctuate over a considerably wider range. The range of summer brood nest temperatures can be exceedingly narrow; Himmer (1927) found daily fluctuations of only 0.6°C, and a monthly range of 33.2–36.0°C, while Simpson (1961) reported temperatures maintained within 0.5°C of 35.0°C. The brood area can be kept at these temperatures over ambient temperature ranges of −40–40°C.

Although the challenge of keeping the nest warm in winter (or cool in summer) is met by a colony-level response, it is the sum of individual worker behaviors which actually regulates nest temperature. Workers can elevate their body temperatures by contracting the flight muscles in the thorax without moving the wings, thereby generating heat (Esch 1960, 1964a; Roth, 1965; Esch and Bastian, 1968; Bastian and Esch, 1970). At rest, a worker's temperature is approximately that of the air temperature, but metabolic rate is elevated significantly when active. Metabolic rate is measured as microliters of oxygen consumed per bee per minute; these figures can be highly variable, but have been measured at about 3 for a resting bee, 25 for a grooming worker, 68 for a bee moving occasionally and and vibrating its wings, and 146–460 for a flying worker (Kosmin, Alpatov, and Resnitschenko, 1932; Jongbloed and Wiersma, 1934). A worker warming the nest using thoracic muscle vibrations is probably respiring at a level similar to a flying bee (Bastian and Esch, 1970).

The behavior of individual workers in clusters has been investigated by Esch (1960). He observed that workers in the outer shell hang together with neighboring workers almost motionlessly, with almost no difference in their thoracic and abdominal temperatures. However, shell individuals occasionally move into the center of the cluster, when their thoracic temperatures jump almost 10°C, from 23–27° to 33–37°C, within minutes. They may remain in the cluster center for up to 12 hr, increasing thoracic temperature whenever it begins to fall, and eventually rejoining the shell workers. Thus, the cluster can be thought of as a dynamic system in which heat is generated metabolically in the loose center cluster and retained by the tighter insulating outer shell, with individuals performing both warming and insulative tasks at different times. A similar mechanism of thoracic heating probably operates in brood incubation at temperatures above the 18°C clustering limit but below the 35°C optimum for brood rearing. Under these conditions a thin layer of workers may cover the brood area, generating just enough heat to warm the brood. The mechanism for brood warming might not involve any special perception of brood temperature; workers could regulate their own body temperature to near 35°C by adjusting their rate of heat production (Cahill and Lustick, 1976), and in doing so could warm the brood area to its optimum temperature. But more recent results suggest that workers adjust their metabolic rate in response to the temperature of the capped brood rather than to their own body temperature (Kronenberg and Heller, 1982).

As ambient temperatures rise, nest cooling becomes increasingly important, particularly when brood is present. Temperatures above 36°C for any appreciable time are harmful to brood, and excesses of only 1–2°C can cause developmental abnormalities and death (Himmer, 1927; reviewed by Jay, 1963a). The range of ambient temperatures through which colonies can be kept cool are as impressive as the range of winter temperatures at which colonies generate heat. In one experiment a colony was placed in full sunlight on a lava field in southern Italy, and although the outside temperature rose to 60°C, the maximum internal hive temperature never exceeded 36°C (Lindauer, 1954). To cool the nest, workers employ a number of tactics, the simplest being adult dispersal through the colony. As internal nest temperature rises, workers begin to ventilate the nest by fanning, evaporating water, and even partially evacuating the nest under extreme conditions.

Ventilation generally begins at or before a nest reaches a temperature of 36°C, with fanning workers lining up in chains facing the same direction throughout the brood nest. Other workers at the entrance face inward and fan, producing cooling air currents and suction, which draws the warm air out of the nest. If fanning does not prove adequate, workers can further cool the nest by water evaporation. The efficacy of this form of air conditioning was demonstrated by Chadwick (1931) in California on a hot June day when temperatures reached 48°C. During the day, when water could be collected, the bees were able to regulate nest temperature, but at night a hot breeze from the desert raised the air temperature to 38°C; when the colony exhausted its water supply, many of the wax combs melted.

Water is used for cooling by spreading it through the nest in puddles on capped cells, as a thin covering over open cells, or as hanging droplets. Workers fanning over the cells increase the cooling evaporative power of the disbursed water. More rapid evaporation can be induced by what has been named "tongue-lashing" behavior, in which workers hanging over brood cells repeatedly extend and contract their probosces, pressing a drop of water from their mouths into a thin film which can evaporate quickly (Lindauer, 1954; Kiechle, 1961). A similar mechanism is used by workers in the field to cool down when flying at overly high ambient temperatures (Heinrich, 1979b, 1980a,b). If further nest cooling is needed, many of the workers will leave the nest and cluster outside, presumably reducing heat generated by their metabolism and also providing more room in the nest for ventilation and water evaporation.

Such precise control over nest temperatures under widely variable ambient temperature conditions requires mechanisms by which workers can not only perceive temperature but also determine which of many thermoregulatory tactics would be the most appropriate. The ability of workers to resolve minor temperature differences was clearly demonstrated by Heran (1952), who showed that bees can detect temperature differences as small as 0.25°C, probably with their antennal thermoreceptors. However, the colony-level mechanisms which integrate individual behaviors are not as well understood, and hypotheses to explain how colonies regulate thermoregulation can be divided into two groups, the "superorganism" concept, as exemplified by Southwick and Mugaas (1971), and the "individual worker" hypothesis, as proposed by Heinrich (1985). According to the superorganism concept, workers do not act independently but rather subordinate their individual needs and responses to a coordinated, colony-level response. According to the individual worker hypothesis, workers essentially behave selfishly, maintaining their own body temperature at optimal levels for different seasons and conditions, and colony-level thermoregulation is simply an advantageous side effect of individual behaviors.

Differentiation between these hypotheses is difficult, since the same worker behaviors can be used to support either concept. A good example is the communication system used to regulate water collection for cooling (Lindauer, 1954). When a small central comb area is artificially heated, the colony's foragers are stimulated to collect water without having contacted the warmed area. The mechanism for this stimulation involves the reception of foragers by nest bees: when water is needed for cooling, foragers returning with either water or very dilute nectar are met enthusiastically, while those with concentrated nectar have difficulty unloading it; this stimulates additional water foraging. When the water needs of the colony have been met, foragers have difficulty unloading water, which stimulates them to collect less water. Unloading times of up to 60 sec inspire more water collection, while those longer than 60 sec discourage it; water unloading times greater than 180 sec almost eliminate water foraging. The superorganism concept argues that such behavior involves a colony-level response to overheating, whereas the individual worker hypothesis argues that the workers who accept water simply use it to cool themselves and stop accepting water when they are sufficiently cool. A second example is fanning at the nest entrance. Temperature of air currents at the entrance is not the stimulus used to evoke fanning; it is the velocity and vibration frequency of air streams produced by fanning workers inside that are perceived by workers at the entrance, who then begin fanning. Again, this appears to be a colony-level phenomenon, but Heinrich has suggested that these fanning individuals are simply cooling themselves. A third example is brood warming; the superorganism concept states that workers regulate the temperature of the brood while in the brood area, whereas the individual worker hypothesis claims that workers are keeping their own temperature constant at the desired 35°C, and the warming of the brood is only an advantageous by-product of those individual behaviors.

The most compelling argument for individual thermoregulation involves the behavior of workers in swarms (Heinrich, 1981a,b,c). Swarm clusters have problems similar to clusters in nests and appear to respond to them in similar ways, that is, workers alternate between the core and the outer mantle, and the central workers generate the heat while the outer workers provide insulation. The core is maintained at about 36°C while the mantle is 15–21°C, a differential similar to that found in colony clusters, although at higher temperatures. When ambient temperatures get too warm, the swarm cluster spreads out, creating channels through which cooling air can move passively or actively by fanning (Fig. 7.2). Heinrich initially interpreted these observations as compatible with some degree of central control, but a number of experiments designed to test this concept were inconsistent with the superorganism theory. First, at low temperatures there seemed to be little or no exchange of workers between the swarm core and mantle, and thermoregulation was unaltered by preventing movement of workers between the core and mantle. Second, exposing workers from one swarm in a warm or cold chamber to pheromones or sounds from another in a chamber under the opposite conditions did not change thermoregulation, indicating that thermal needs were not being communicated by these mechanisms. Third, the total metabolic rate of swarms was not different than that expected from a composite of bees respiring at their resting rates, indicating that higher metabolism of the workers in the core may be a passive function of the higher respiratory rates due to crowding rather than an active increase in metabolism designed to keep the mantle bees warm. This last observation is an important one, since Southwick and Mugaas proposed that a colony cluster thermoregulates itself to maintain a constant cluster surface temperature rather than as individual workers trying to keep warm. Further work on thermoregulation in colonies is required to delineate these two hypotheses, but it is entirely possible that both colony-level and individual mechanisms result in proper thermoregulation.

Colonies also regulate other environmental conditions, particularly levels of carbon dioxide. Colonies respond to high levels of carbon dioxide by fanning (Hazelhoff, 1941), and carbon dioxide levels are maintained between 0.1 and 4.3% (Seeley, 1974), well within the range of detection which honey bees are capable of (Lacher, 1964). Higher levels of carbon dioxide may inhibit certain enzyme reactions and cause problems of water retention, so maintenance of lower levels is an important part of nest homeostasis.

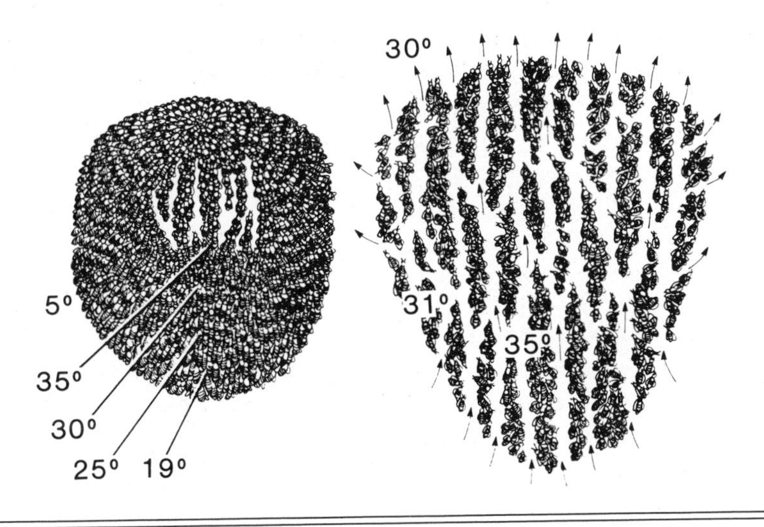

Fig. 7.2

Temperature (°C) regulation in a swarm cluster. Under cooler conditions workers cluster tightly, with little internal ventilation, and maintain a core temperature higher than at the surface. Under warmer conditions the cluster spreads out to cool the center, and surface and core temperatures are more uniform. (Redrawn from Heinrich, 1981.)

Worker Behavior in Queenless Colonies

Queen loss is one of the most serious events which can occur in a colony; at best, it results in a prolonged period without brood rearing, at worst, in colony death, since workers are not always successful in replacing lost queens. Bee researchers have been interested in emergency queen rearing because the contrast between worker behavior in these unusual situations and in queenright colonies provides insights into normal colony functioning and colony responses to stress. Worker behaviors differ considerably depending on whether the queen is lost when brood is present or absent in the colony. When brood is present, the colony attempts to rear a new queen, but when brood is absent, queen replacement is difficult or impossible. Queen loss as discussed below does not include queen replacement due to supersedure or reproductive swarming (these topics are considered in Chapter 11).

The events following queen loss in colonies with brood present have been most recently described by Fletcher and Tribe (1977a,b), Winston (1979b), Punnett and Winston (1983), and Fell and Morse (1984). Queens can die suddenly for a number of reasons, most commonly disease or predator attack, although the frequency of such events in natural colonies has never been determined. It is clear, however, that the highest priority of colonies following queen loss is to rear a new queen from the eggs or larvae present in the colony, since failure to rear a new queen usually results in colony death. Workers are able to detect the absence of their queen within about 10 hr (Seeley, 1979) and then begin a series of events which, if successful, result in the presence of a new, mated queen within about 4 weeks.

The earliest changes in worker behavior following queen loss are an apparent nervousness, aggressiveness, and increased walking throughout the colony; a "roaring" sound can be heard upon opening queenless colonies due to increased scenting behavior. The first tangible evidence of queen loss is the construction of queen cups in the colony, followed by queen rearing (Fig. 7.3). Workers generally begin queen cell construction directly over eggs or larvae, and the brood in those cells are given special feeding and reared as queens. But queen cells can be constructed anywhere in the nest, and workers occasionally move brood into empty cells to rear them as queens. Most queen cells are constructed during the first 12–48 hr after queen loss, although queen rearing may begin in some cells constructed up to 9 days after queen loss, and colonies continue to build queen cups and attempt queen rearing for up to 12 days. Such late queen rearing presumably uses brood which were newly laid eggs at the time of queen loss and which may have had slow development times caused by lower temperatures at the periphery of the brood nest. Colonies attempt on average to rear about 20 queens, although mortality of the queen brood is high, and only an average of 12–15 queens are usually reared successfully to adulthood. Workers generally begin queen rearing with larvae less than 2 days old, probably because larvae that young will produce better queens than older larvae. Some mistakes are made as well; workers will start queen rearing with 4- or 5-day-old worker larvae and even drone larvae, although brood reared in such cases are generally aborted. The priority given to queen rearing is evident from the high mortality rate of nonqueen brood, averaging 40–50% following queen loss.

Once mature queens begin to emerge in colonies, they either mate or swarm; the incidence of swarming in emergency queen situations can be up to 100% of colonies, with some colonies producing two or even three swarms with virgin queens. The advantage of swarming behavior is not immediately apparent, since swarming further weakens a colony already weakened by high brood mortality and a long broodless period. One possible explanation is that workers are not able to perceive differences between queen loss due to death and queen loss due to reproductive swarming. Also, if the cause of

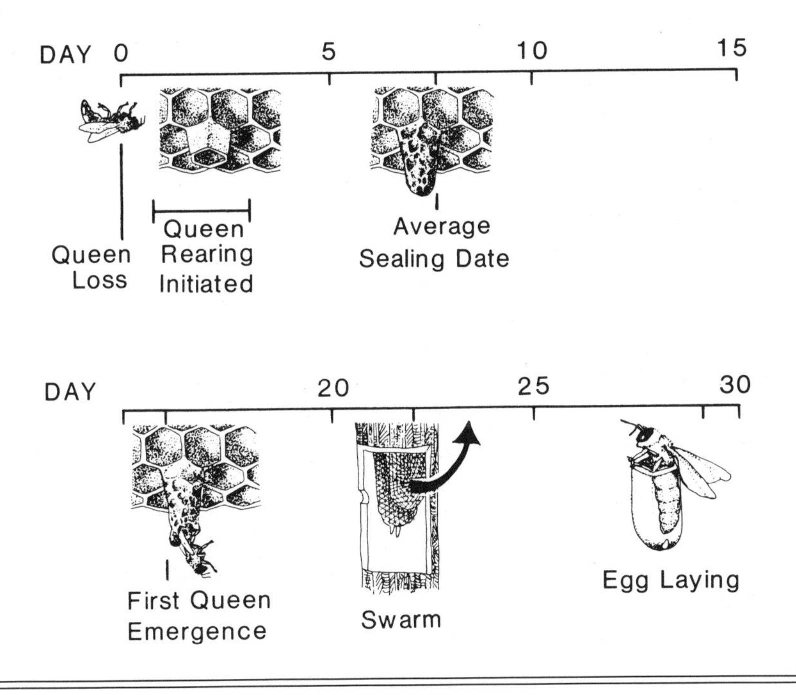

The ontogeny of events following queen loss, showing average times for the initiation of queen rearing, sealing queen cells, first queen emergence, queen-loss swarming, and egg laying by the new queen.

queenlessness is likely to recur (for example, disease, predation), it may be advantageous for weak colonies to produce swarms which colonize new localities. Still, the relative fitness of queen-loss swarms as opposed to other types of swarms has yet to be investigated. At any rate, eventually one queen kills all of the other emerged queens in the colony, and she then mates and starts egg laying. The entire process from queen loss to egg laying by a new queen takes an average of 29 days, and most colonies succeed in rearing a new queen.

There are some racial differences in responses to emergency queen situations between temperate- and tropical-evolved bees, most notably in the timing of new queen production and the frequency of brood movement and queen-loss swarming. For example, colonies of Africanized bees in South America tend to use older larvae to initiate queen rearing, which has the advantage of shortening the queenless period but the disadvantage of reduced queen viability. African bees, from the area where the South American bees originated, use larvae of ages similar to those used by European races, suggesting that selection for a shorter queenless period for Africanized bees may have occurred in South America. The mean duration of queenlessness in

African and Africanized colonies is only 23 days, as compared to 29 days in European colonies, partly because of the shorter queen-rearing time for Africanized bees and partly because of the shorter time between the establishment of the new colony queen and her initiation of egg laying. Also, Africanized workers move an average of 47% of brood reared as queens into empty cells, compared to about 4% moved by European bees. The advantage of this behavior is not clear, but it does result in more queen rearing at the periphery of colonies of Africanized bees during emergency queen-rearing situations. Finally, a higher proportion of Africanized colonies swarm during emergency queen situations than of European colonies, and Africanized colonies usually produce two or three swarms as opposed to the one swarm more commonly issuing from European colonies. This higher frequency of swarming may be an effect of the higher tendency of tropical-evolved colonies to swarm under any conditions.

The events following queen loss are quite different in colonies without brood, or in colonies with brood which fail to requeen themselves. Workers can begin egg laying under these conditions, generally producing only drone eggs, although in some cases parthenogenetic female eggs are produced which can develop into queens (see Chapter 4). Many of the workers in queenright colonies may have the potential to begin laying eggs, but the presence of a queen in the colony and of brood inhibits worker egg laying. Ovary development and oogenesis by workers are inhibited by the queen pheromone 9-keto-decenoic acid as well as other queen-produced substances (Butler and Fairey, 1963), and some as yet unidentified odors from brood are also thought to inhibit the development of laying workers (Pain, 1961; Jay, 1972). In the queen's absence both ovaries and mandibular glands of workers enlarge (Costa-Leonardo, 1985).

The latency period for worker egg laying after a colony loses its queen varies between different races, with European races averaging 23–30 days and African races only 5–10 days before workers begin laying eggs. Laying workers of European races also lay fewer eggs than African-evolved bees (Ruttner and Hesse, 1981). Once workers have begun laying eggs, colonies are characterized by aggressiveness and fighting between workers and generally will not accept a new queen (Sakagami, 1954). The fate of such a hopelessly queenless colony is eventual death, since only male brood are produced.

Occasionally colonies are found which have constructed only drone comb and have only worker-laid drones in them; these presumably originated from swarms which lost their queens in transit and developed laying workers. Sometimes one laying worker in a colony will develop a higher level of queenliness than the others and will be treated as a queen (Sakagami, 1958). This laying worker is called a false queen and is attended by a retinue of workers which feed and lick her like a queen. False queens have slightly swollen abdomens and seem to inhibit ovarian development and oviposition of other workers, probably because of increased production of mandibular gland substances identical to those produced by real queens (Velthuis, Verheijen, and Gottenbos, 1965; Crewe and Velthuis, 1980).

Perhaps the most interesting situation involving laying workers is found in the Cape bee *A. m. capensis*, a race of honey bee located only at the southwestern tip of South Africa. This peculiar race of bee is distinguished from all other honey bee races in the high frequency of parthenogenetic female eggs produced by laying workers following queen loss; in fact, very few drone eggs are laid. This characteristic was reported as early as 1912 (Onions, 1912, 1914; Jack, 1916), but was not widely believed until Anderson (1963) conclusively demonstrated that laying *capensis* workers did indeed produce female brood. In addition, the workers of *A. m. capensis* differ from other races in having larger although nonfunctional spermathecae (Ruttner, 1977), high numbers of ovarioles (Anderson, 1963; Ruttner, 1977), and queen substances in the mandibular glands (Ruttner, Koeniger, and Veith, 1976; Hemmling, Koeniger, and Veith, 1979; Crewe, 1982).

The loss of a *capensis* queen is followed by extreme fighting in the colony, much more pronounced than among other races, possibly because these laying workers have the potential for producing queens, and thus their offspring could inherit the colony. The first worker-laid eggs are found within 4–8 days following queen loss, and workers begin egg laying even if brood is present when the colony loses its queen. Some of the brood used for queen rearing originates from queen-laid eggs, but some queens originate from parthenogenetic worker-laid eggs as well. Colonies may remain queenless but producing brood for up to 4 months, all female brood being produced by laying workers; eventually a new queen is reared.

One of the least-understood aspects of honey bee biology is the absence of a high frequency of diploid egg production by workers among all races except for *A. m. capensis*. Production of female brood by workers should be highly favored by selection, since it would ensure colony survival following queen loss, yet this trait is common only in *capensis*. Female egg laying by workers has not spread from the Cape bee to adjacent populations of *A. m. scutellata*, although extensive hybridization between the two races appears to be occurring (Moritz and Kauhausen, 1984). It has been suggested that the aggressive *scutellata* workers may kill hybrids which have laying worker characteristics, since these workers appear more queenlike than regular workers; this might explain the failure of parthenogenetic worker production to spread beyond the Cape bee (Fletcher, personal communication). After examining worker characteristics in hybrid zones where *capensis* and *scutellata* overlap, Moritz and Kauhausen (1984) have concluded that *capensis* is in danger of extinction as a result of extensive hybridization with *scutellata* and pressure from commercial beekeeping. Loss of this unique race of bees would indeed be tragic, partly because of its potential importance as a genetic reserve of unique traits, and partly because we still do not understand why its characteristics have remained isolated at the tip of South Africa.

8

The Chemical World of Honey Bees

All animals must communicate with members of their own species, for mate location if nothing else. Similarly, orientation is a necessary function, at least to find food and locate the most favorable environment for survival. In social animals these functions are particularly complex and sophisticated. Even those insect species with only limited social behavior, for example, can communicate alarm, find mates, establish territories, and orient back to an established nest site following foraging trips. In the more advanced social insects a wide array of sensory inputs are routinely processed and acted upon by individuals, but with an integration into colony performance much greater than the sum of individual behaviors.

Even among the highly social insects, honey bees stand out for their extraordinary communication and orientation abilities. They not only possess one of the most intricate chemical communication systems in the social insects, but also have evolved a dance language unparalleled in its ability to communicate the location of food resources and nest sites. In addition, honey bees use visual, auditory, and magnetic cues to round out their sensory impressions of the world, and they integrate all of these sensory abilities into rapid and effective colony-level responses to both danger and opportunities.

Pheromones

Pheromones are chemicals used for communication among members of a species. In honey bees these odors can be produced by workers, queens, and possibly drones, and are known to function for mating, alarm, defense, orientation, colony recognition, and integration of colony activities (Table 8.1). At least 18 chemicals have been identified which function as pheromones, and it has been estimated that 18 additional substances must exist to describe fully pheromone-based activities (Pain, 1973). Our understanding of pheromones expanded dramatically in the 1960's, with the advent of instrumentation capable of sampling and identifying these substances, and in-

Table 8.1 Pheromones produced by honey bees (? indicates no known gland or chemical)

Pheromone	Gland	Chemicals	Functions
Worker-produced			
Nasonov	Nasonov	Geraniol Nerolic acid Geranic acid (*E*)-citral (*Z*)-citral (*E-E*)-farnesol Nerol	Orientation
Footprint	(Arnhart)	?	Orientation
Forage marking	?	?	Orientation at flowers
Alarm	Mandibular	2-Heptanone	Alarm and defense
Alarm	Sting	Isoamyl acetate 2-nonanol *N*-butyl acetate *N*-hexyl acetate Benzyl acetate Isopentyl alcohol *N*-octyl acetate	Alarm and defense
		(Z)-11-eicosen-1-ol	Alarm and defense orientation
Recognition	?	?	Kin and/or colony recognition
Queen-produced			
Queen substances	Mandibular	9-keto-(*E*)-2-decenoic acid (9ODA) 9-hydroxy-(*E*)-2-decenoic acid (9HDA)	Inhibition of queen rearing Inhibition of worker ovary development Drone attraction Worker attraction to swarms Swarm cluster stabilization Stimulate release of Nasonov pheromone Stimulate worker foraging Queen recognition
Olfactory	Koschevnikov	?	Worker attraction

Table 8.1 continued

Pheromone	Gland	Chemicals	Functions
Tergite	Tergite	?	Drone attraction and copulation Inhibition of worker ovary development Inhibition of queen rearing
Immature	?	?	Inhibition of queen rearing
Footprint	?	?	Inhibition of queen cup construction
Drone-produced			
Marking	Mandibular	?	Marking of congregation areas
Brood-produced			
Brood	?	?	Stimulate foraging Brood recognition Inhibition of worker ovary development
Comb-produced			
Hoarding	?	?	Increase nectar storage

creased again in recent years, with our growing ability to identify substances from single insects and to use sophisticated bioassays to test new compounds. Nevertheless, chemical communication in honey bees remains a largely unexplored world, and each new discovery highlights how little we still know about this complex subject. What is clear, however, is that individual bees are deluged with biologically meaningful odors produced by other bees, flowers, and nest materials, and that sifting through and responding to these odors is a critical aspect of colony functioning.

Worker-Produced Odors

Worker-produced pheromones are known to function for orientation as well as alarm and defense. Considering this short list of functions, there are a surprising number of chemicals involved—16 identified plus others unidentified (Table 8.1; see also Chapter 3 on glands). Some of this chemical overload can be explained by the somewhat different contexts in which each group of chemicals is used, but it is also likely that at least some of these substances may have other undetermined functions.

The pheromones produced by the Nasonov gland are the most well-known orientation odors and are composed of seven volatile chemicals (Table 8.1; Boch and Shearer, 1962, 1964; Weaver, Weaver, and Law, 1964; Butler and Calam, 1969; Pickett et al., 1980). The secretion and functioning of these biosynthetically related terpenoids have been elucidated in a series of experiments at Rothamsted Experimental Station in England (Pickett et al., 1981; Williams, Pickett, and Martin, 1981, 1982; see also references cited therein). The major component in the Nasonov secretion is geraniol, but the most attractive components in both field tests and in the magnitude of antennal electrophysiological responses are (*E*)-citral and geranic acid. This apparent discrepancy in composition and efficacy of Nasonov pheromones can be explained by a highly specific enzyme system in the Nasonov gland which converts the major component, geraniol, into the more attractive (*E*)-citral and geranic acid. Each component, however, contributes to the attractiveness of the blend, and the strongest responses generally are induced by the full mixture of compounds. Newly emerged workers produce very little Nasonov pheromone, but the amounts increase rapidly with age, peaking when the bee is about 28 days old, coincidental with foraging activity.

Workers expose the Nasonov gland and disperse the odors by fanning in a number of orienting situations, including nest entrance finding, forage marking, and swarming. At the entrance, the Nasonov scent is important in guiding into the colony workers such as incoming foragers, workers on orientation flights, and workers disoriented following colony disturbance. The Nasonov dispersing behavior at colony entrances can be induced by many colony odors, among them empty comb, honey, pollen, and propolis, as well as by one of the queen-produced pheromones, 9-HDA (Ferguson and Free, 1981). The Nasonov pheromones themselves induce exposure of the Nasonov gland, which explains the rapid increase in the number of scenting bees that occurs as workers cluster, enter a new nest site, or enter the nest after temporarily being denied access.

Workers also release the Nasonov pheromones when foraging for water, at artificial dishes containing sugar syrup, and, very rarely, at flowers (von Frisch, 1923; von Frisch and Rösch, 1926; Free and Racey, 1966; Free and Williams, 1970, 1972). At water collection sites release of the Nasonov pheromones provides an odor to assist incoming water foragers in orienting to a relatively odorless resource.

Workers can also be induced to fan and release Nasonov scents at artificial food dishes containing sugar concentrations greater than those of most flowers, although Nasonov release diminishes when workers come from colonies with good food storage (Pflumm, 1969; Pflumm et al., 1978; Pflumm and Wilhelm, 1982; Wilhelm and Pflumm, 1983). This suggests that workers can control their scent-making behavior according to a combination of the characteristics of the sugar solution imbibed and the condition of their colony's food supply. But workers are rarely seen scenting at flowers, suggesting that the primary function of forage marking is water collection, and Nasonov pheromones might only be released at flowers with unusually high sugar concentrations or at nests being robbed of their honey.

The Nasonov chemicals are also important for orientation during swarming, including clustering after a swarm issues and orienting to the entrance of the new nest (Morse and Boch, 1971; Ferguson et al., 1979; Free, Ferguson, and Pickett, 1981; Free et al., 1982). In both contexts the Nasonov odors act in concert with queen-produced pheromones in providing cues; presumably, the queen in turn uses the worker odors for orientation. When a swarm cluster forms, orienting workers initially orient visually to small groups of loosely grouped workers already settled on the substrate, but release of Nasonov scent is quickly initiated, facilitating orientation of the remaining workers and the queen. Only five of the Nasonov components are important for cluster formation; nerol and farnesol do not improve the efficacy of the blend. At a new nest site, the first arriving workers begin to scent-fan, again facilitating orientation of the remaining incoming workers and the queen.

The other worker-produced orientation compounds include the as yet unidentified footprint pheromones, (Z)-11-eicosen-1-ol, and a forage-marking pheromone. Footprint pheromones are odors left behind by walking worker bees at the nest entrance and possibly at flowers which attract other incoming workers (Ribbands, 1955; Lecomte, 1956; Butler, Fletcher, and Watler, 1969; Ferguson and Free, 1979; Williams, Pickett, and Martin, 1981). These odors are deposited by the feet when the Nasonov gland is not exposed and may be produced in the tarsal Arnhart glands (Chauvin, 1962). However, extractions of chemicals from various parts of worker's bodies, especially the thorax, are more attractive than leg extractions, and it is likely that the footprint substance is produced elsewhere on the bee and only deposited by the feet. The presence of footprint pheromones enhances the attractiveness of the Nasonov odors at artificial food sites, and both the airborne Nasonov pheromones and the substrate footprint pheromones are used for nest entrance orientation. Thus, these two groups of chemicals act synergistically, at least at the nest entrance and when used for forage marking.

A sting-produced compound, (Z)-11-eicosen-1-ol, has been shown to be an alarm substance, but it also attracts foraging workers (Pickett, Williams, and Martin, 1982; Free et al., 1982). Unlike footprint pheromones, however, it does not enhance the attractiveness of Nasonov scent. Another group of odors, called forage-marking pheromone by Ferguson and Free (1979), induces foragers trained to a food source to alight; these odors may be produced on the dorsal surface of the abdomen. Thus, it appears that there may be at least four different pheromone-based orientation systems produced by honey bee workers: Nasonov, footprint, (Z)-11, and the forage-marking pheromone. Some of these may actually be the same pheromones; for example, the forage-marking pheromone could be any of the other three, and footprint pheromones may include sting components. Also, the role of these odors in other orientation contexts, such as swarming, has not been investigated, and the forage-marking substances have been shown to act at artificial food dishes but not at flowers. The activity of these compounds needs considerable clarification, but it is at least clear that workers produce numerous attractive compounds used for orientation functions.

Workers also produce numerous odors used for alarm and colony defense (Table 8.1). One of these substances, 2-heptanone, is produced in the mandibular glands of workers; all of the other alarm chemicals are produced in the sting, probably in setose membranes at the base of the sting lancets. As a group, these compounds alert workers to danger, lower the threshold of sensitivity for the attack reaction, and assist workers in orienting to spots on attackers which have already been stung. But alarm pheromones are not sufficient to elicit full defensive behavior. Workers exposed to alarm substances placed on stationary objects at the nest entrance become agitated, assume characteristic aggressive postures, dash toward and crowd around the object, but generally do not sting unless the object is moving (Free, 1961a; Maschwitz, 1964a,b; Free and Simpson, 1968; Boch, Shearer, and Petrasovits, 1970; Gary, 1974).

The alarm substances, when presented separately, do not all elicit the full defensive responses, and some are more active than others. The weakest alarm odor is 2-heptanone, first isolated by Shearer and Boch (1965). This chemical does elicit aggressive responses in guard bees at colony entrances, but 20 to 70 times the amount of 2-heptanone may be required to produce a reaction comparable to that of the sting-produced isoamyl acetate (Boch, Shearer, and Petrasovits, 1970). The primary function of 2-heptanone may be to repel robber bees and possibly other enemies at nest entrances, since it is strongly repellent to foraging workers (Butler, 1966; Simpson, 1966).

The first sting compound to be identified as an alarm substance was isoamyl acetate, also called isopentyl acetate by some authors (Boch, Shearer, and Stone, 1972). This substance can be dispersed by workers opening their sting chamber and fanning, but is also released when workers sting, presumably rupturing the membranes which produce it (Ghent and Gary, 1962; Maschwitz, 1964a). Isoamyl acetate not only serves as an alarm substance, but also attracts workers to a previously stung spot, thereby increasing the intensity of a colony's alarm response. This behavior has been noted by generations of beekeepers familiar with the "rule" that the probability of being stung increases greatly after an initial sting is received, and that smelling the sweet, banana-like odor of isoamyl acetate is a good signal to close the hive and come back another day. The response to isoamyl acetate depends on concentration; increasing amounts of this alarm odor produce quicker, longer-lasting, and stronger reactions (Collins and Rothenbuhler, 1978). The amount of 2-heptanone present in worker stings and mandibular glands also is correlated with the intensity of aggressive responses, at least at the colony entrance (Kerr et al., 1974).

More recently, a number of additional sting-produced compounds have been proposed as alarm pheromones (Table 8.1; Blum et al., 1978; Collins and Blum, 1982, 1983; Pickett, Williams, and Martin, 1982). Although all of these compounds elicit similar alarm reactions, they vary in the intensity of worker responses following exposure; 2-nonanol, isoamyl acetate, and (Z)-11-eicosen-1-ol are the most active sting-produced compounds, and 2-heptanone showed similar activity in a standardized laboratory test.

Why do honey bee workers produce such an impressive array of compounds for alarm and orientation? Although it is not unusual for insects to produce pheromones with multiple components, honey bees do produce an unusual number of substances for relatively few functions. One possible reason is that each of the chemicals may be involved in slightly different responses. For example, some of the sting components may be more effective at eliciting general alarm response, while others may be involved in orientation to targets. For Nasonov odors, use of all seven components at the nest entrance increases orientation ability, although there are some differences in response intensity when components are used separately, and two of the components are not important for swarm clustering orientation. It appears that some of the orientation compounds may be more important as entrance markers, whereas others may function to mark forage. It is also possible that a limited number of sensory receptors are capable of perceiving each compound, and use of multiple components could increase the potential level of alarm response.

Another explanation for the large number of worker-produced pheromones may be that some of these compounds are involved in functions still to be determined. Among the evidence suggesting this interpretation are differences in the ages when workers produce at least one group of pheromones, 2-nonanol and three chemically related compounds. Young workers produce little or no isoamyl acetate, and maximal production is not reached until workers are 15–25 days old (Boch and Shearer, 1966; Allan, Winston, and Slessor, unpublished observations). The pattern for 2-nonanol production is quite different; almost none of this or related compounds are found in workers before 20 days of age, and production by workers older than this age is erratic, with most workers sampled not producing any appreciable amount of these pheromones (Fig. 8.1). Furthermore, although samples of workers inside hives, fanning workers, guards, and foragers all show high levels of isoamyl acetate, foragers produce significantly more 2-nonanol than do the other sampled categories (Fig. 8.2). These data imply that 2-nonanol may have some function associated with foraging, thereby differing from the other alarm odors.

The responses of workers to pheromones depend on such factors as age, temperature, and humidity. Newly emerged workers show little behavioral or neurophysiological response to alarm pheromones and queen-produced odors, but strong responses are apparent by the time workers are 5–10 days old (Collins, 1980; Masson and Arnold, 1984; Allan, Winston, and Slessor, unpublished observations) (Fig. 8.3). The strongest responses may occur when workers are approximately 28 days old; there appears to be a dip in response intensity immediately prior to that time, possibly associated with the change from nest to field duties, and another drop in workers 36 days of age and older, possibly due to senescence. Responses to alarm pheromone are more intense with higher temperatures and, to a lesser extent, higher relative humidities (Collins, 1981).

Queen-Produced Odors

In workers, there are a multitude of chemicals involved in very few functions; in queens, there are few identified chemicals involved in a multitude of functions (see Table 8.1). The major known queen pheromones are two acids produced in the mandibular glands, 9-keto-(*E*)-2-decenoic acid (also called 9-oxodecenoic acid, abbreviated as 9ODA) and 9-hydroxy-(*E*)-2-decenoic acid (9HDA). The first of these

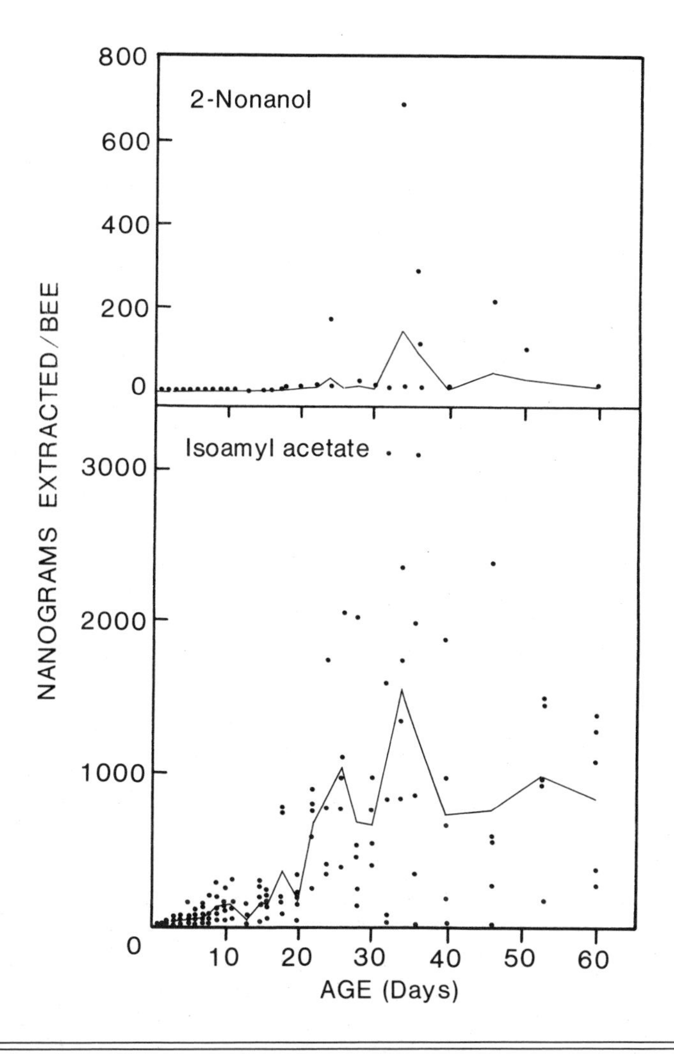

Fig. 8.1

The amount of two pheromones, isoamyl acetate and 2-nonanol, extracted from worker bees of various ages. Each point represents a single worker, and the lines show the means for each age. (Data from Allan, Winston, and Slessor, unpublished observations.)

to be identified, 9ODA, has been called "queen substance," and this chemical has received the most research attention. Its discovery began with the observation that the removal of laying queens from colonies results in rapid changes in worker behavior, followed by queen rearing and eventually worker ovary development. But workers exposed to the queen's mandibular glands, which inhibit queen rearing and worker ovary development, behave as if a queen were present (Butler, 1959). Eventually, 9ODA and 9HDA were identified as the

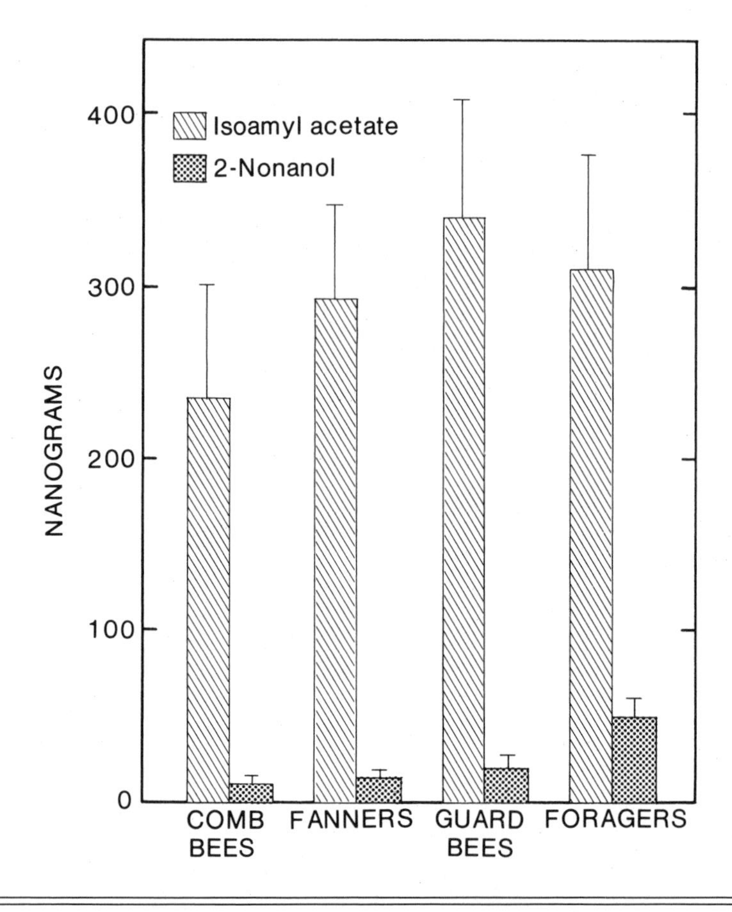

Fig. 8.2

The relationship between task and the amount of pheromones produced by workers. Comb bees were removed from brood areas inside colonies, fanners and guard bees were picked from the entrance while performing these activities, and foragers were returning to hives with engorged honey stomachs or carrying pollen. (Data from Allan, Winston, and Slessor, unpublished observations.)

active mandibular gland components responsible for these and other effects (Barbier and Lederer, 1960; Barbier, Lederer, and Nomura, 1960; Callow and Johnston, 1960; Barbier and Hugel, 1961; Butler, Callow, and Johnston, 1961). Both 9ODA and 9HDA are highly specific in the relationship between their chemical configuration and their activity; closely related compounds and isomers of these queen substances show reduced or no biological activity (Callow, Chapman, and Paton, 1964; Doolittle, Blum, and Boch, 1970; Winston et al., 1982). Other pheromones may be produced by the abdominal Koschevnikov and dermal tergite glands and by immature queens, but although the biological activity of substances from these sources has been demonstrated, the active chemicals have yet to be identified (Renner and Baumann, 1964; Butler and Simpson, 1965; Renner and Vierling, 1977; Velthuis, 1970; Boch, 1979; Free and Ferguson, 1982).

The amounts of 9ODA and 9HDA produced by queens, and therefore the effects of the queen's secretions, depend on queen age, mating status, time of day, and season. Of 9ODA, virgin queens less than 2 days old produce on average only 7 μg, whereas virgins 5–10 days old produce 108–133 μg, and mated laying queens less than 18 months old produce 100–200 μg. Thus, the secretion of 9ODA is a function of queen age rather than a result of mating and/or egg laying. Older laying queens show reduced 9ODA production, and this reduced output may be associated with queen supersedure by work-

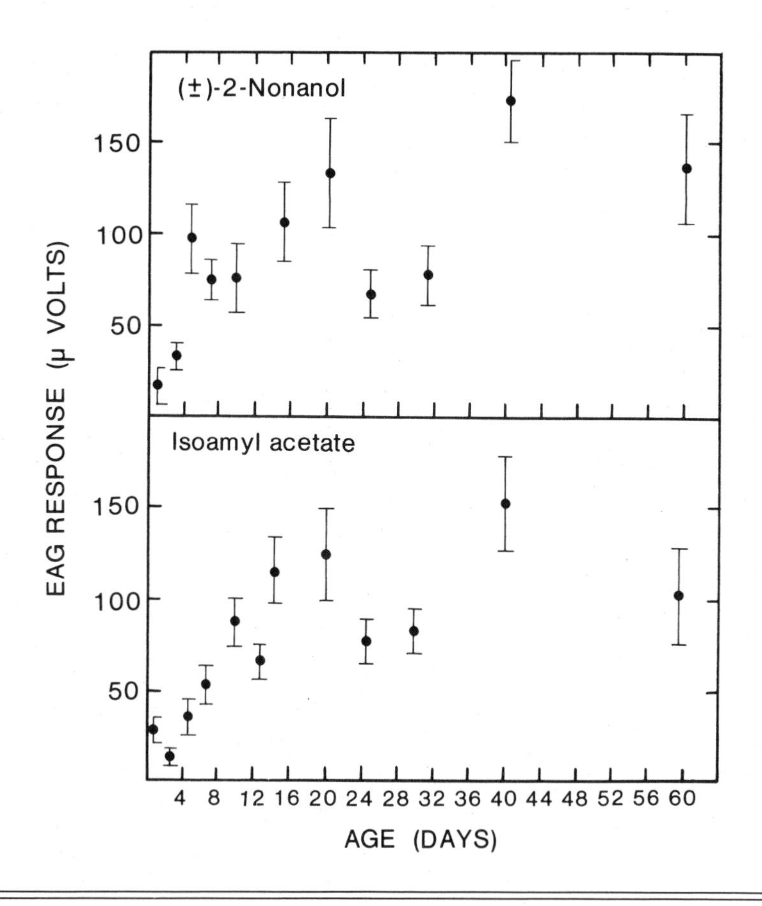

Fig. 8.3

Electroantennogram responses of workers of various ages to 2-nonanol and isoamyl acetate. (Data from Allan, Winston, and Slessor, unpublished observations.)

ers (Butler and Paton, 1962; Callow, Chapman, and Paton, 1964; Pain, Barbier, and Roger, 1967; Shearer et al., 1970). The reduced biological activity of extracts from young virgin queens compared to laying queens is consistent with their lower 9ODA production (Butler, 1957a, 1960). The 9HDA pheromone is produced in considerably lower quantities than 9ODA by laying queens, about 5 μg per queen (Callow, Chapman, and Paton, 1964). Virgin queens show diminished production of both 9ODA and 9HDA in late summer and early winter, when colonies are less active, and peak production of 9ODA during afternoons, their prime mating time (Pain, Roger, and Theurkauff, 1972, 1974; Pain and Roger, 1978).

Both 9ODA and 9HDA as well as the unidentified pheromones have been shown to have numerous biological functions, including inhibition of queen rearing and swarming, prevention of worker ovary development, attraction of drones for mating, attraction of workers to swarms, swarm cluster stabilization, stimulation of Nasonov pheromone release, induction of worker foraging, and queen recognition. However, although it is likely that queen-produced pheromones are involved to some extent in these functions, their effects are difficult to evaluate for a number of reasons. Research of these substances has utilized a wide range of techniques using field, within-colony, and laboratory caged bee assays. In addition, different concentrations, isomers, and combinations of chemicals and extracts from bees have been used, and in some cases poor syntheses of these pheromones have resulted in the wrong chemicals or mixtures being produced. Also, factors such as worker and queen age and physiological condition are often not standardized, and most colony functions are also controlled by nonchemical behavioral mechanisms in addition to pheromones. Nevertheless, queen substances are involved in at least some of the following six functions.

INHIBITION OF QUEEN REARING AND SWARMING

One of the first functions of 9ODA and 9HDA to be demonstrated by scientists was the inhibition of queen rearing, which prevents reproduction by swarming or supersedure. The earliest research (reviewed by Butler, 1959, 1960) showed that substances produced in the queen's mandibular glands prevent workers from rearing queens, and that these substances are more active in glands from mated queens than from virgins. It was also shown that the amount of queen substances necessary to inhibit a colony from rearing queens increases with the number of workers in a colony, and naturally superseded or swarming queens contain less queen substance than mated and laying queens. Later work confirmed that both 9ODA and 9HDA act together to suppress queen rearing (Butler, 1961; Butler and Paton, 1962; Butler and Callow, 1968); unidentified substances from the abdominal tergite glands may have a similar effect (Velthuis, 1970). Interestingly, workers which have visited cells containing queen pupae move through the colony and make antennal contacts with other workers, indicative of pheromone transfer, and the presence of queen pupae inhibits queen cell production; these effects are not found for queen eggs or larvae (Boch, 1979; Free and Ferguson, 1982). Thus, immature queen pupae produce substances which prevent additional queen rearing, possibly but not necessarily 9ODA and/or 9HDA.

The footprint pheromone produced in the queen's tarsal glands also may be involved in the suppression of the first stage of queen rearing associated with swarming, cup construction. A queen walking along a glass surface deposits an oily, colorless trail at a rate of 0.8 mg of material/hr, and this material evidently originates in the tarsal Arnhart glands. When a combination of mandibular and tarsal gland extracts were applied to the bottom edges of comb in overcrowded colonies, the construction of queen cups was inhibited (Lensky and Slabezki, 1981). Neither of these secretions affected queen cup construction when applied separately. Thus, colony overcrowding may prevent the queen from moving along the bottom of combs and depositing the footprint pheromone. This, in combination with the mandibular gland secretion, may inhibit the construction of queen cups and thereby delay or prevent swarming. The active ingredient, however, has not yet been isolated, and it is not clear whether there is any relationship between this pheromone and the worker footprint pheromone.

PREVENTION OF WORKER OVARY DEVELOPMENT

Numerous studies have demonstrated that queen-produced substances prevent worker ovary development and egg laying (Voogd, 1955; Butler, 1957b, 1959; Verheijen-Voogd, 1959; Pain, 1961; Butler and Fairey, 1963; Velthuis 1970, 1972). Only 9ODA has been investigated for this effect, and although it does inhibit worker ovary development, the inhibition is not as effective as the presence of a mated queen. Thus, other pheromones or behaviors associated with mated queens may be required for full inhibition, including substances from the queen's abdominal tergite glands in addition to brood-produced pheromones that are important for worker ovary suppression (Jay, 1970; Velthuis, 1970).

ATTRACTION OF DRONES FOR MATING

The attraction of drones to queens for mating was initially demonstrated in response to mandibular gland extracts, and subsequently

9ODA and to a lesser extent 9HDA were found to be the active components (Gary, 1962; Butler and Fairey, 1964). Drones may be attracted by queen pheromones at distances up to 60 m, and 9ODA may function at close range as an aphrodisiac to stimulate the mounting of queens by drones. Unidentified pheromones of the dermal tergite glands have been found to attract drones at distances less than 30 cm and increase their copulation activity (Renner and Vierling, 1977), and it is possible that attractive pheromones from the Koschevnikov glands may be released when the queen's sting chamber is opened prior to mating (Butler and Simpson, 1965; Gary, 1974).

ATTRACTION TO SWARMS AND SWARM CLUSTER STABILIZATION

The queen's mandibular gland secretions have three swarm-associated functions: they attract workers to the cluster, stabilize the cluster, and aid in swarm movement to a new nest site (Morse, 1963; Simpson and Riedel, 1963; Butler, Callow, and Chapman, 1964; Butler and Simpson, 1967; Morse and Boch, 1971; Avitabile, Morse, and Boch, 1975; Winston et al., 1982). When workers issue from a nest in a swarm, they are attracted to the queen wherever she has alighted; 9ODA appears to be more important for attracting flying workers, whereas 9HDA seems to function more to stimulate alighting and clustering. Once the cluster has formed, both pheromones, but 9HDA in particular, help to stabilize the cluster and prevent the workers from becoming restless and leaving the cluster prior to the swarm's movement to a new nest site. When the swarm lifts off to travel to the new nest, workers sense the presence of the queen through the 9ODA odor, and the swarm moves as a group only if the queen is present. For both cluster formation and swarm movement, Nasonov pheromones act in concert with the queen-produced substances.

STIMULATION OF NASONOV PHEROMONE RELEASE AND WORKER FORAGING

Exposure to 9HDA stimulates workers to release Nasonov pheromones at nest entrances, but 9ODA has no similar effect (Ferguson and Free, 1981). This difference suggests that 9HDA rather than 9ODA may stimulate Nasonov release by workers which have recently settled on swarm clusters. Foraging for nectar can also be stimulated by the presence of a laying queen, queen extract, or 9ODA alone (Jaycox, 1970a), although there are certainly many additional factors besides pheromones which are important for the induction of foraging.

QUEEN ATTRACTIVENESS AND RECOGNITION

Finally, queen substances are important in worker recognition of and attraction to their queen. There is little doubt that 9ODA and probably 9HDA attract workers and allow them to differentiate between fellow workers and queens; substances from the abdomen also have the same effect (Butler and Simpson, 1965; Velthuis, 1970; Simpson, 1979).

It is clear that the two known queen-produced pheromones, 9ODA and 9HDA, as well as some unidentified compounds, exert considerable influence on worker behaviors and colony functions. There has been considerable controversy, however, over queen pheromones, which is not surprising considering the complexity of pheromonal effects and the lack of identified compounds to work with. Three much-researched aspects of queen pheromones—the role of 9HDA, the relative importance of mandibular substances as opposed to other unidentified pheromones, and the mode of transmission and action of queen substances—are the topics of the remainder of this section.

While 9HDA has been shown to be active either alone or synergistically with 9ODA in the five functions discussed above, other studies have shown little or no activity of 9HDA in attracting workers to swarms (Morse and Boch, 1971), inhibiting queen rearing (Boch and Lensky, 1976), and attracting drones (Blum et al., 1971; Boch, Shearer, and Young, 1975). The conflicting nature of these results has been compounded by the disclosure that 9HDA used in some of the experiments was of questionable purity (Boch, Shearer, and Young, 1975). But the 9-hydroxyl function of 9HDA is chiral (Fig. 8.4), and it is possible that the enantiomer used might influence the behavioral response. We investigated this problem using swarm clustering behavior as a bioassay (Winston et al., 1982), and demonstrated that the R(−) form was significantly more effective in retarding swarm dispersal than the S(+) form, and, further, that the 9HDA produced by queens is predominantly (85%) the more active R(−) form (Slessor et al., 1985). Thus, at least some of the differences in activity attributed to 9HDA by various researchers appear to have been due to the form of the pheromone used. The importance of identifying and synthesizing the active isomer of these queen substances was emphasized by Doolittle, Blum, and Boch (1970), who found that the cis (*Z*) isomer of 9ODA was not attractive to drones but that the trans (*E*) isomer was.

A second controversial subject in pheromone research is the relative efficacies of mandibular and other queen substances. One of the more interesting approaches to this problem was devised by Gary (1961a), who developed a technique to remove the mandib-

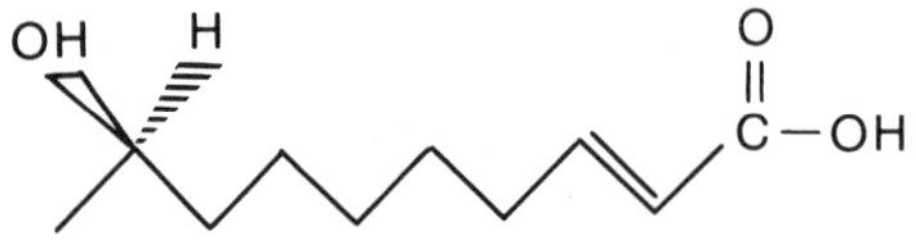

Fig. 8.4

The enantiomers of 9-hydroxy-(*E*)-2-decenoic acid, showing the two chiral forms of the molecule.

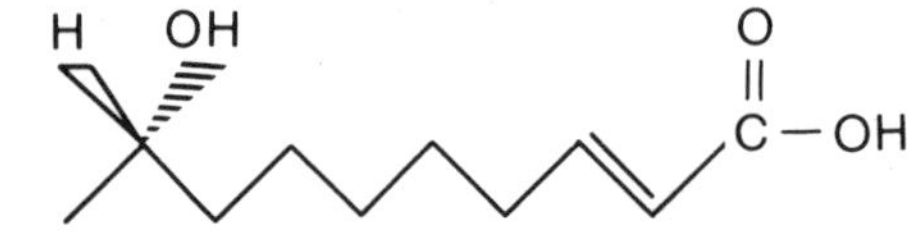

ular glands from queens; demandibulated queens can survive and, if mated, lay eggs for many months. In some studies queens without mandibular glands lost approximately 85% of their attractiveness to workers when introduced to colonies in cages and in swarms (Gary, 1961b; Velthuis and van Es, 1964). This finding suggests that the mandibular secretions are the most important component attracting workers to queens. But when mated queens whose mandibular glands had been removed were introduced into small colonies by Velthuis and van Es (1964) and Velthuis (1970), there was a surprising result. In these experiments the queens were accepted by colonies and functioned normally for at least 3 months. Colonies with these demandibulated queens formed normal retinues around their queen and did not attempt to rear new queens, and workers showed no increase in ovary development. The active pheromones responsible for these normal behaviors were found to originate in the abdominal glands. These results should not be interpreted to mean that 9ODA and 9HDA do not influence these colony functions; their importance has been clearly demonstrated in many other studies. Rather, as Velthuis (1970) properly concludes, mandibular and other pheromones have closely linked effects on colonies; in some situations one group of pheromones may suffice to effect complete control over a colony function, and in other situations pheromonal effects may be additive. Thus, the behavioral context in which pheromones act is critical in understanding their effects.

Another area of intense interest is the mode of transmission and action of queen pheromones. Some transmission mechanism besides direct worker-queen contacts must operate, since workers not directly in contact with the queen nevertheless show the effects of pheromonal activity. For example, in one experiment two groups of workers were separated from each other by a mesh screen through which they could touch each other but not pass. One group was given a caged queen while the other remained queenless, but neither the queenright nor the queenless group showed any ovary development (Verheijen-Voogd, 1959).

Three transmission modes have been proposed: transmission by food exchange between workers, volatile transmission, and transmission between the body surfaces of workers, particularly during antennal contacts. Of these, only the volatile transmission mode has been conclusively demonstrated. Although neither of the identified mandibular secretions are particularly volatile, both workers and drones outside the nest are attracted to queens, queen extracts, and 9ODA and/or 9HDA; workers are attracted during swarming, and drones for mating. In both cases only airborne odors can explain the attraction.

Regarding the other two hypotheses, most evidence supports surface transport rather than transmission by food exchange. Since workers do exchange liquid materials frequently, the food exchange hypothesis is an attractive one (reviewed by Gary, 1974, and Michener, 1974). Nixon and Ribbands (1952) have shown that radioactively labeled sugar syrup can be distributed to most workers in a colony within hours. Although queen extracts presented in drinking water showed some inhibitory effects on worker behavior, extracts mixed with sugar syrup higher than 5–10% in concentration showed no effects (Verheijen-Voogd, 1959; Erp, 1960). The earlier reports may actually have recorded surface transport of pheromones picked up during feeding.

In contrast, there is abundant evidence for the surface transport hypothesis. First, radioactively labeled 9ODA was found to move on immobilized workers from one body part to another, both externally and internally, through translocation, although some movement of radioactive material may have been due to labile tritium rather than 9ODA (Butler et al., 1974). Second, a body part or other surface that has been in contact with a queen becomes attractive to worker bees (Verheijen-Voogd, 1959; Velthuis, 1972). Third, there are specific olfactory receptor cells in the antennae specialized for 9ODA perception (Kaissling and Renner, 1968). Fourth, behavior of workers attending queens and following queen attendance strongly supports the idea of surface pheromone transport (Verheijen-Voogd, 1959; Velthuis, 1972; Seeley, 1979; Ferguson and Free, 1980). Workers in the queen's retinue touch her with their antennae, particularly, and tongue (Fig. 8.5). Most frequently, the antennal tips are lightly and quickly brushed over the queen, although at times the entire anten-

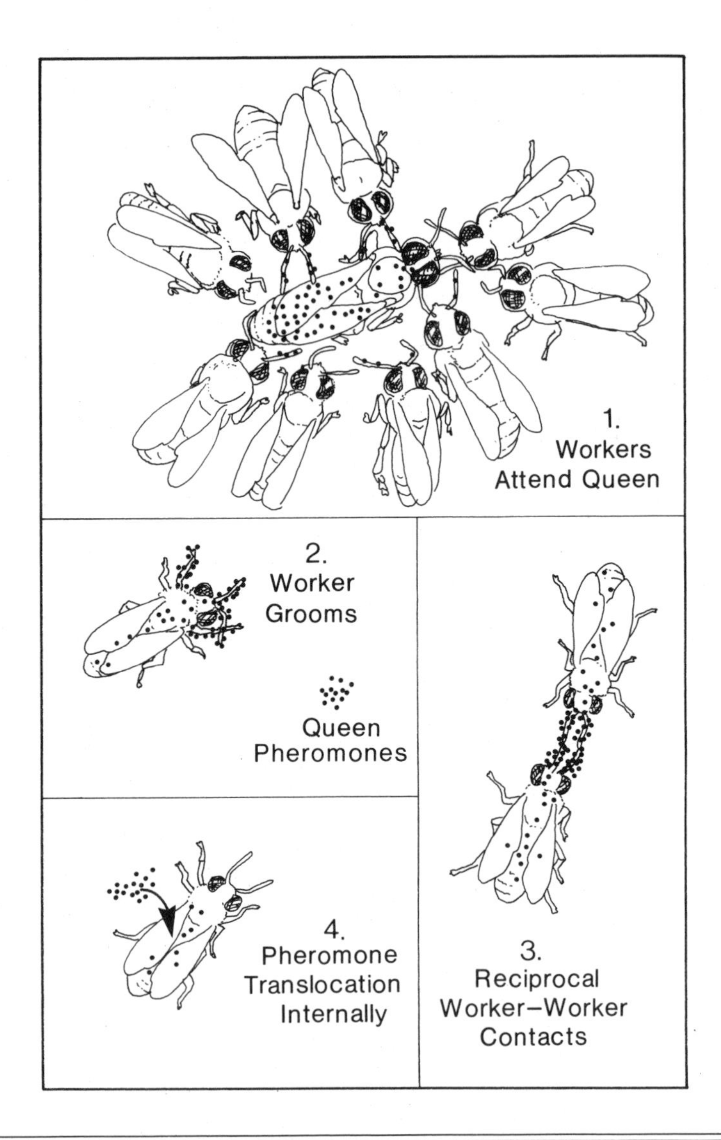

Fig. 8.5

The hypothesized movement of queen-produced pheromones by worker messenger bees to other workers, through surface transport and internal translocation.

nae move over the queen's body in a flurry of activity. Following these contacts, workers generally groom their antennae with their forelegs, presumably passing pheromones between the antennae, tongue, and forelegs. Most significant, workers which have contacted queens in this manner move through the colony making reciprocated antennal contacts with other workers at high rates for approximately 30 min.

A typical worker will exchange nestmate antennations with an average of 56 other workers during this time but only exchange food twice. Workers tend to perform these activities between 3 and 9 days of age; although workers of these ages do not specialize in queen contacts, they are particularly attracted to queen substances (Pham, Roger, and Pain, 1982). Seeley (1979) proposed the name "messenger bees" for these workers, and he and the others cited above have concluded that pheromones are distributed over the queen's body by her grooming, picked up by direct contacts between workers and the queen, and then transported throughout the colony by these messenger bees.

Queen substances probably act on the worker's hormonal system when the pheromones are translocated internally from the worker's body surface. The injection of 9ODA partly suppressed ovary development in one experiment (Butler and Fairey, 1963), probably owing to inhibited growth of the corpora allata, the gland that secretes hormones responsible for the stimulation of ovary development (Lüscher and Walker, 1963; Gast, 1967). The growth of neurosecretory cells may also be inhibited by queen substances (Gast, 1967). Numerous studies have shown that queen substances show a gradual decline in their activity over a 30-min period, with a half-life of 15–20 min (Velthuis, 1972; Juska, 1978; Seeley, 1979; Juska, Seeley, and Velthuis, 1981). The mechanism for this decline in effectiveness may involve volatilization, but the known queen pheromones are not particularly volatile, suggesting that some enzymatic deactivation mechanism may metabolize queen substances into inactive forms (Pain and Barbier, 1981). This pheromone deactivation is quite important for colony functions, since it can serve as a signal to alter worker behavior in response to changes in colony conditions.

Other Pheromones

In addition to queen and adult worker-produced pheromones, there is some evidence that drones and comb may produce pheromones, and good evidence that brood produces pheromones. In addition, bee and nest odors interact to provide recognition cues which workers can use to recognize their own colony and queen.

A recent study demonstrated that mandibular gland secretions were attractive to drones flying in congregation areas (Lensky et al., 1985). The attractive compounds are evidently synthesized by secretory cells during the first 9 days of the drone's life and then stored until use. This work supports the earlier suggestion of Gerig (1971, 1972) that drone heads are attractive to flying drones. However, the function of these attractive compounds remains unknown; since drone attraction to other drones has no known biological function, these pheromones may be more important in attracting queens to congregation areas.

Volatiles from empty comb have been found to increase the rate of nectar storage. This may function to stimulate foraging for nectar in colonies which have space for more honey storage (Rinderer, 1981).

Brood odors appear to have at least two effects on adult workers: they stimulate foraging and prevent ovary development. The presence of brood stimulates foraging for pollen in particular (see reviews by Free, 1967b, and Scott, 1986), and an extract of worker larvae has the same stimulatory effect (Jaycox, 1970b). Worker ovary development is inhibited by either unsealed or sealed worker brood, in either queenright or queenless situations (Kropacova and Haslbachova, 1970, 1971; Jay, 1972). There may be both a volatile and a contact component by which workers recognize the presence of brood (Jay, 1972; Free and Winder, 1983); one contact pheromone may be glyceryl-1,2-dioleate-3-palmitate (Koeniger and Veith, 1983). The purpose of both worker brood and queens inhibiting adult worker ovary development is to permit the onset of worker egg laying only in hopelessly queenless situations. If a colony loses its queen, the presence of worker brood allows the colony to rear a new queen, and thus it is not advantageous for laying workers to develop. In contrast, a lack of brood means that the only reproductive possibility left to a colony is for laying workers to develop, and in fact this development takes only 14 days in colonies from which both worker brood and the queen have been removed, as compared to 30 days in colonies with brood but without the queen.

A final role of pheromones is in kin and colony recognition, whereby workers recognize their own queen, colony, and in some cases differentiate between sisters and half-sisters. The earliest evidence for nestmate recognition comes from observations on guard bee behavior at nest entrances. In one study almost all nestmates were admitted to colonies, whereas less than 40% of non-nestmates were admitted (Butler and Free, 1952). For centuries beekeepers have observed that a foreign queen introduced to a colony is quickly killed by the workers, unless she is protected by some type of cage for a few days until the workers become accustomed to her. Queen recognition has also been demonstrated using swarms in which workers given a choice between clustering with their own or a foreign queen generally choose their own (Boch and Morse, 1974, 1979; Ambrose, Morse, and Boch, 1979).

The potential cues for nestmate recognition could be visual, auditory, or olfactory, although the first two are unlikely, since recognition can occur in the dark interior of a nest, and bees have a very limited sense of hearing. The use of odors in recognition was first demonstrated by Kalmus and Ribbands (1952), who showed that foragers from two colonies trained to adjoining sugar dishes were preferentially attracted to the dish visited by members of their own colony, and that this attraction was due to some odors carried by workers. All subsequent research has confirmed the olfactory nature of nestmate recognition (reviewed by Breed, 1985).

There are two potential sources of recognition odors, the bees themselves and environmentally acquired odors from flowers, propolis, or other sources, both of which appear to be involved. Renner (1960) demonstrated that recognition could be manipulated by applying external odors, and Boch and Morse (1981) showed that recognition of queens could be manipulated by applying artificial scents. Bee-produced odors as well are involved in nestmate recognition, as has been demonstrated by experiments in which all environmental sources were kept constant, nestmate and non-nestmate workers or queens were removed from colonies prior to or at emergence, and then these individuals were reintroduced to their own or foreign colonies some days later (Breed, 1981, 1983b). Significantly more nestmate workers and queens were accepted upon introduction than non-nestmate, indicating that bees produce internal cues by which workers can discriminate between their colony members and foreign workers or queens. Even more remarkable are recent studies (to be discussed in Chapter 12) demonstrating that workers can discriminate between sister and half-sister workers and queens, in both the larval and adult stages. While all of these nestmate recognition capabilities are undoubtedly based on odors, none of the relevant chemicals or ratios of chemicals have yet been identified. Recognition cues may even involve subtle differences in ratios or concentrations of pheromones, rather than different chemicals, since workers can detect small differences in concentration of even a single compound (Kramer, 1976).

In summary, it is quite evident that pheromones exert considerable influence on honey bee behavior, and that we have only begun to understand these effects. Now that we are aware of how much more there is to learn about pheromones and able to take advantage of recent advances in the techniques used by chemists to analyze minute quantities of odors from single insects, the influence of pheromones on bee behavior will certainly be one of the principal research areas in honey bee biology for years to come.

9

Communication and Orientation

As social insects, bees are able to integrate their activities so that the sum of colony functions is much greater than what individuals could achieve independently. For this integration to occur in a meaningful way, individuals must be able to communicate, particularly to inform nestmates about available resources outside the nest. Communication, coupled with orientation mechanisms which can function over long distances, allows recruitment and exploitation of resources and is one of the foundations of social insect colonies.

Probably no subject in all of animal behavior has received more attention than the communication and orientation mechanisms of honey bees. The earliest and simplest observations of such behavior were that within minutes of a honey bee worker's discovering a dish of honey, one, then a few, then tens of honey bee workers would appear, even if the nearest colony was kilometers away. Clearly, the first scouts to discover a resource are able to communicate its location to other workers back at the nest. The nature of the mechanisms involved in this communication and in orientation to and from the nest has taken over a half-century to discover and still provides one of the most significant ongoing research topics in animal behavior.

The core of recruitment in honey bees is a dance language, by which remarkably precise information about the distance, direction, and quality of resources is communicated between individual workers performing various dances inside the nest or in swarms. Closely related to that language are the orientation mechanisms by which workers can leave and return to their nest; thus, research on communication has been closely linked to studies of orientation. These studies largely originated with the brilliant work of Karl von Frisch and his students and have involved some of the most refined, extensive, and carefully designed research in all of science. This work has been most thoroughly reviewed in von Frisch's classic book, *The Dance Language and Orientation of Bees* (1967a), which even at that time was able to incorporate close to 800 references. Other recent reviews have been done by Lindauer (1961), von Frisch (1967b), Michener

(1974), Gould (1976), and Dyer and Gould (1983); complete references for the topics discussed below can be found in those sources.

Dance Language

Although the discovery of honey bee dance language seems relatively straightforward with hindsight, it actually required a long series of deceptively simple experiments which eventually demonstrated that honey bees possess this abstract, symbolic method of communication. The existence of some form of communication was first demonstrated by Maeterlinck at the turn of the century in an experiment in which he let a forager find a food dish with honey in it and then return to the colony. When the forager left the colony for another flight, she was caught and not allowed to return to the food source. Nevertheless, other workers quickly appeared at the dish, evidently having found it by some information passed on in the colony by the original worker. Using similar techniques in glass-walled observation hives, von Frisch and others observed that a returning forager frequently performed a dance which would be followed by other workers in the hive; the attending workers would run after the dancer, making antennal contacts with her body and sampling regurgitated food. From these experiments it initially appeared that recruits learned the odor of a food source from the scout bee, and that this olfactory mechanism explained the dance communication.

This explanation was not convincing, however, since it was found that workers could be recruited to flowers over distances of many kilometers, even flying downwind, and odor alone could not explain this phenomenon. In one experiment von Frisch set out numerous scented dishes of sugar syrup at various distances and directions from a colony. Since only some of these were discovered, it became evident that odor alone was not sufficient to explain recruitment. Over many years of observations and experiments, von Frisch and his colleagues gradually determined that information about distance, direction, and quality of resources was being transmitted during the dances, and that in fact there were a number of dance types used in different situations. From these studies and subsequent work by other researchers, three basic dances have been found: a round dance, a waggle dance, and a dorsoventral abdominal vibrating dance (DVAV). In addition, other dances involving jostling, buzzing, shaking, and trembling have been seen, but these are still poorly understood.

The round dance is used primarily for recruitment to resources, the waggle dance for resource recruitment and transmitting information about the location of new nest sites, and the DVAV dance for stimulating foraging and controlling queen emergence and the timing of swarming during colony reproduction. All of these dances can be performed either on the comb surface inside colonies or outside on the surface of swarms. Inside colonies dances are usually performed near the entrance, and potential recruits can be found waiting there for scouts to inform them about the location of resources.

To study these dances only three methodological components are required: an observation hive to observe dances in, food dishes containing sugar syrup outside the colony, and some type of paint or glue-on numbers to mark workers with. There is perhaps no better proof of the existence of a dance language than the fact that, once the language is learned, a trained observer can use these tools to read the language and find out exactly where the dancing workers are going.

THE ROUND DANCE

The round dance is the simplest dance and does not communicate precise distance or direction information. Rather, it simply informs workers that there is a resource within close proximity to the nest, less than 15 m away. In performing this dance, an incoming worker which has discovered a nearby food source first exchanges nectar with workers inside the nest. Then she performs the round dance, being closely followed and antennated by attending bees (Fig. 9.1). In this dance the dancer repeatedly makes small circles, reversing and going the opposite direction after every 1 or 2 revolutions, and sometimes more frequently. Up to 20 of these reversals can occur, with dances lasting for only seconds or up to minutes. Often, food is then exchanged again between the dancer and nest bees, and then dancing may resume. The dancer may then leave the nest on another foraging trip, while the recruits clean themselves, take some honey for energy on their flight, and then exit the colony. In one group of experiments von Frisch showed that of 174 workers which had contacted a dancer, 155 (89%) were found at a food dish within 5 min, indicating the rapid recruitment potential of the round dance.

There is no evidence, however, that the round dance transmits information concerning the exact location of the food source, and it seems that exiting recruits must search the immediate vicinity of the nest to find the resource. Evidently, recruits fly in increasing circles from the nest, and orient to the food source using the odors passed on during dancing. This can be demonstrated by training workers to a sugar solution scented with a strong odor such as anise, citronella, or peppermint. Once recruits appear the scent can be changed, and new recruits trained to the old odor have difficulty finding the sugar dish, whereas recruits to the new odor can find it but not dishes with the original scent. Workers at food dishes may also release Nasonov

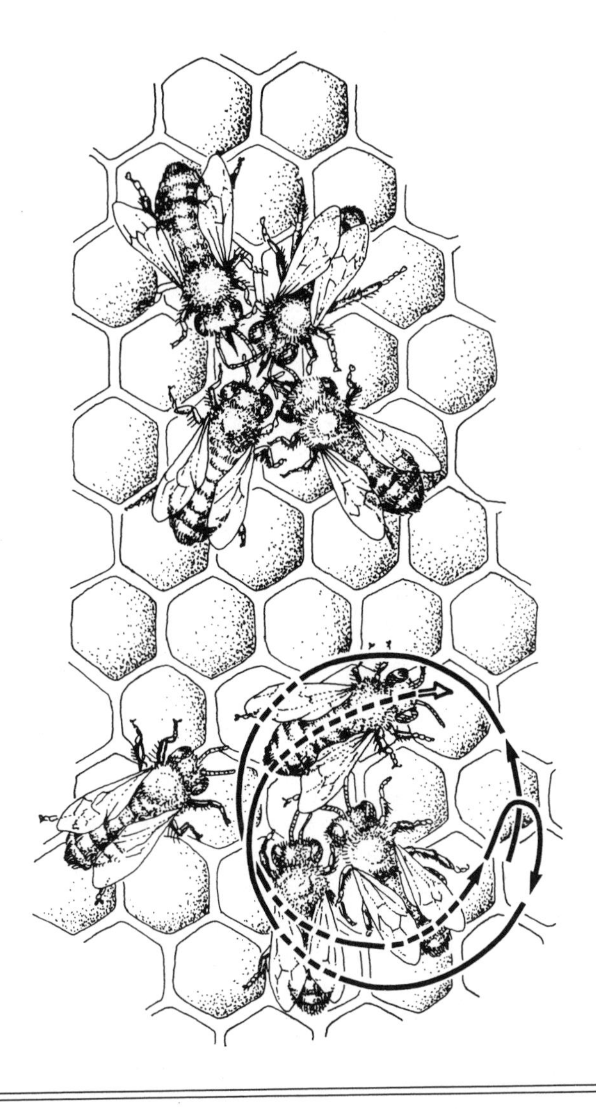

Fig. 9.1

The round dance indicating a resource close to the nest. In the top group of workers a returning forager is exchanging regurgitated nectar with potential recruits; the bottom group is performing the round dance, which is being followed by recruits. (Redrawn from von Frisch, 1967a.)

scent to attract incoming recuits, although this pheromone release is not common on flowers.

The profitability of food rewards can also be communicated through the round dance, so that more concentrated sugar solutions elicit more vigorous and long-lasting dances. A worker that has discovered a sugar source, for example, is more likely to dance upon return for solutions of higher concentration and will dance for longer periods with increasing sugar concentrations as well. The dances also exhibit more "vigor" in that abdominal vibrations are much more pronounced during dances for better resources. The efficacy of this increased dance time and vigor can be shown by the number of workers recruited; for a sugar concentration of only 0.5 molal, 10 workers were recruited in one 30-min observation period, whereas close to twice that number were recruited to a 2-molal solution. Interestingly, the sugar concentrations that elicit the most vigorous dances change during the season, depending on scarcity of nectar in the field. Lindauer (1949) demonstrated that when nectar was abundant, a high concentration of sugar was required to initiate dancing, whereas during a period of nectar scarcity very low sugar concentrations would stimulate the same dancing level.

Workers which have discovered pollen sources also use the round dance to stimulate recruits, in much the same way as for nectar sources. Incoming workers laden with pollen are antennated closely by potential recruits, and their dances are also followed. Pollen odors are transmitted during these exchanges (Hopkins, Jevans, and Boch, 1969), and attending workers are quickly recruited to pollen-bearing flowers of the same type as those visited by the dancer. But it is not clear how the vigor of dances relates to the quality or quantity of available pollen, since the mechanisms whereby bees evaluate pollen sources are not as obvious as for nectar.

While the round dance is used for resources within 15 m of the nest, there are transition dances for resources between 25 and 100 m by which the round dance gradually changes to the waggle dance (Fig. 9.2). Different races of bees use different types of transition dances; the *carnica* race shows a fairly direct transition to a figure-eight pattern, while other races show transitions via a sickle dance. The distance at which the transition is initiated also depends on a colony's racial origin; the Egyptian bee *A. m. fasciata* begins the transition for resources located only 3 m away, while *A. m. ligustica* and *A. m. carnica* begin the transition at 8 and 15 m, respectively.

THE WAGGLE DANCE

Honey bees use the waggle dance to communicate information about the distance, direction, and quality of resources at distances greater than about 100 m from the nest. It has also been called the tail-wagging, or figure-eight, dance, since workers shake their abdomens during some of its "steps," and its characteristic pattern of movement is in the configuration of a figure-eight. In a typical waggle dance the bee runs straight ahead for a short distance, emphasizing its movements by shaking the body vigorously from side to side at a rate of

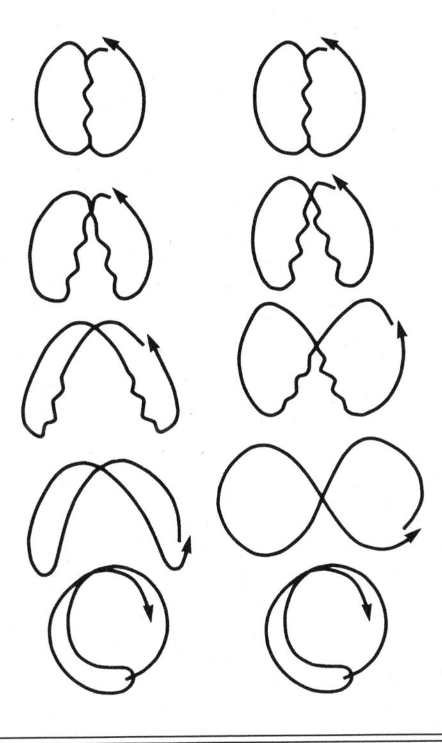

Fig. 9.2

Transition dances from the round to the waggle dance. On the left are the sickle-shaped transition dances used by most bee races, and on the right are the direct transition dances used by *A. m. carnica.* (Redrawn from von Frisch, 1967a.)

about 13–15 times/sec (Fig. 9.3). The abdomen is given the most emphasis during this wagging behavior, and a buzzing sound is given off by muscular and skeletal vibrations. At the end of each straight run the bee turns in one direction and makes a semicircular turn back to the starting point, followed by another straight run and a semicircular turn in the opposite direction. As in the round dance, the waggle dance is punctuated by the dancer stopping and distributing food from its honey stomach to nearby workers, and the dance itself is closely attended by a retinue of workers with extended antennae. The dance followers may produce a squeaking sound of short duration, lasting 0.1–0.2 sec, which has been called a "begging signal" and causes the dancer to halt and exchange food with the bee that squeaked (Esch, 1964b; Michelsen, Kirchner, and Lindauer, 1986). Both nectar and pollen collectors dance in the same manner.

The remarkable aspect of this dance is the way information is translated from the abstract dance language into terms which the attend-

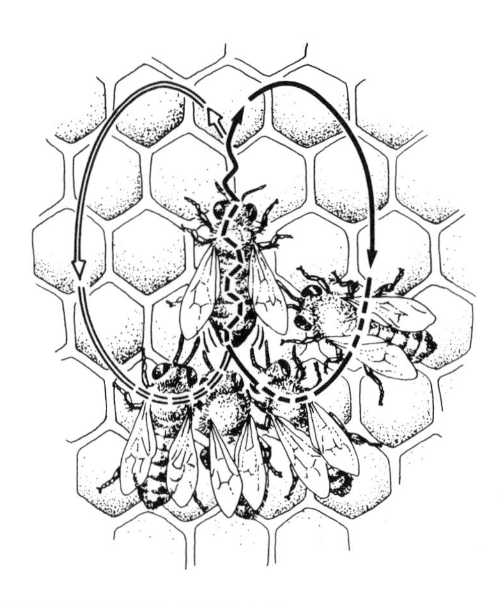

Fig. 9.3

The waggle, or figure-eight, dance, being followed by potential recruits. (Redrawn from von Frisch, 1967a.)

ing workers can read and use to locate resources. For distance information, a number of interdependent factors are evaluated by the workers to determine how far away the resource is, including the length of the straight run in comb cell diameters, the length of the straight run in millimeters, the duration of waggling and buzzing, the number of circuits per unit dance time, and to a lesser extent the average distance of dances from the nest entrance. Dance characteristics most indicative of distance are the dance tempo, measured as the number of circuits per 15 sec, and the duration of waggling and buzzing during the straight run (Fig. 9.4). The tempo of the dance slows down with greater distance, and the time spent in the straight run increases for resources further away. Evidently the duration of the buzzing vibration is as important as the time spent in the straight waggle portion of the dance, and the antennae of workers following the dancers may be as important for perceiving the vibrations produced by the dancers as they are for odor perception (Esch, 1961; Esch, Esch, and Kerr, 1965).

Another notable aspect of distance communication via the waggle dance is that foragers signal not the absolute distance to a food source, but rather the amount of energy expended in getting there. This has been demonstrated in a number of ways. For example, putting small weights or drag-producing flaps on foragers induces dances to greater distances than the actual location of the food source

(Schifferer, 1952), and dances when there is a headwind or the source is uphill also indicate a greater distance. When foragers are forced to walk rather than fly, thereby expending more energy, a much greater distance is also indicated (Bisetzky, 1957). The amount of energy to be expended on a flight can be predicted by recruits from the dances they follow; as the distance to resources increases, more honey is taken into the crop prior to flight.

It is important to recognize, however, that distance communication is not perfect, and all recruits do not find resources on their first trip. Depending on the circumstances and type of experiment, recruits fly to within 2–10% of the distance of a resource. Some of this imprecision is due to variability in dance tempo between individuals, particularly since tempo slows down with the age of the worker, so that older workers indicate greater distances than they did when they

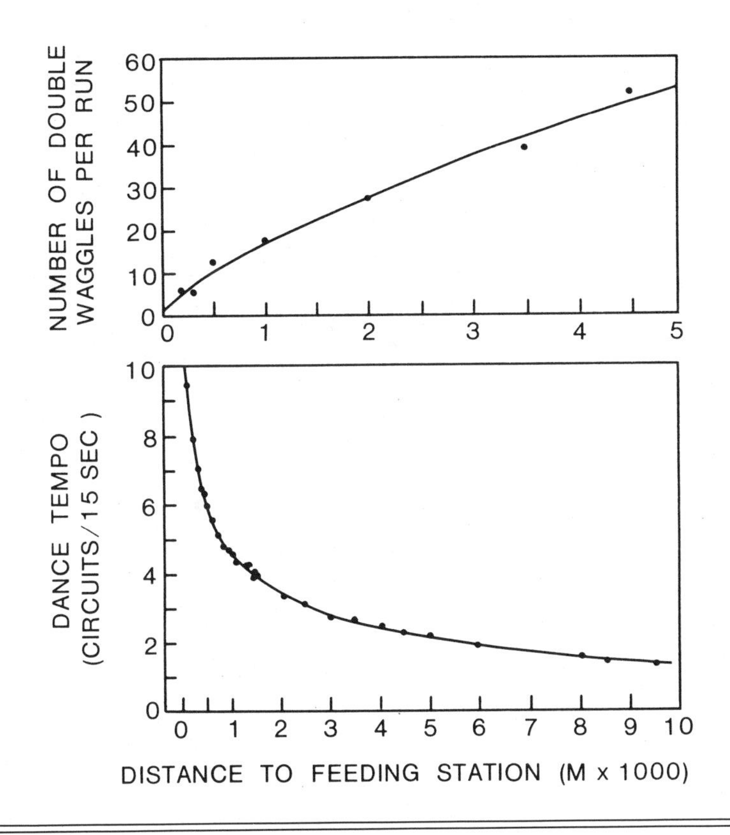

Fig. 9.4

The relationship between distance to a food source and two dance characteristics used by recruits to "read" the dance, the number of waggles per run and the dance tempo. (Redrawn from von Frisch, 1967a, and Michener, 1974.)

were younger. In addition, there are dialects among bees, with different races showing slightly different dances to the same food source. Environmental factors may influence dance tempo, thus the slight increase noted with higher temperatures. Part of these communication problems are resolved by recruits following dances from a number of foragers before exiting the colony, evidently summing the information from all the dances to compute an average distance. Nevertheless, in spite of the imperfections of the dance language, many recruits do get close enough to the resource to find it by visual and olfactory orientation.

The same components of the dance which indicate distance also provide information about the quality of the resource. This can easily be seen in an observation hive, where some dances seem to lack vigor and are poorly attended, while others are vivacious and eagerly followed by potential recruits. The principal dance attributes which communicate resource quality are the lateral extent of waggling, the total number of cycles, and the intensity of the buzzing vibrations; all of these characteristics increase for better resources. Furthermore, a dancing bee with a desirable load is "encouraged" by her nestmates to dance vigorously by their enthusiastic and rapid reception of her crop contents during food exchanges. The round dance is similarly influenced by profitability, with the number of reversals in the dance being correlated with the quality of nectar rewards (Waddington, 1982). The profitability of food is judged by the sweetness and purity of the sugar solution, the ease of securing the food—including distance to it, weather, and floral handling time—and the type and amount of floral odors. In addition, colony conditions dictate in part the vivacity of the dance; under famine conditions, workers dance more intensely and are recruited more easily to lower-quality resources than workers in a well-provisioned colony.

The other remarkable aspect of the waggle dance is that it can transmit information about the direction to a food source in addition to its distance and quality. The simplest method for communicating direction can be seen when workers are forced to dance on a horizontal surface. In that case the direction of the straight waggle run is directly toward the food source, with an average of less than 1° angular deviation from the proper direction. Followers of the dance then simply fly off in the direction indicated. But most dances performed by honey bees are done on the vertical surfaces of comb or swarms, so that the horizontal direction in the field must somehow be translated from the vertical orientation of the dance.

Horizontal direction is indicated by transposing the solar angle into the gravitational angle (Fig. 9.5). For example, if the food is in the exact direction of the sun, the dance will be performed so that the

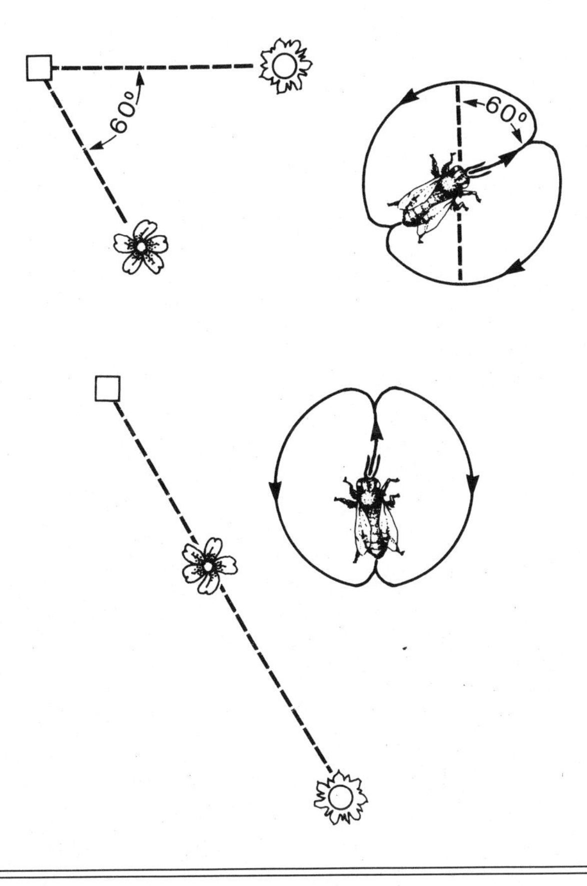

Fig. 9.5

Orientations of two waggle dances on vertical comb relative to the direction to a floral food source. *Top:* the flower is 60° to the right of the sun, and the dance is 60° to the right of vertical. *Bottom:* the flower is in the same direction as the sun, and the bee is dancing straight up.

straight run is straight up. If the food is 60° to the right of the sun, the straight run part of the dance will be done at 60° to the right of vertical. Workers use specialized organs at the base of the neck to detect the gravitational angle of the dance, and they are able to translate that information into the azimuth of the sun when they exit the colony. If these organs are eliminated by careful dissection, workers are no longer able to orient properly to a food source.

Bees face the same problems any navigator does; the sun moves or can be obscured behind clouds suddenly during the day, and weather conditions such as crosswinds must be taken into account for accurate orientation. Bees evidently use a time correction factor coupled with some computation of where the sun should be, so that they correct for the sun's movement during the time they are out on their flight and generally can return to the colony if the sun should become obscured. They also compensate for drift due to crosswinds by turning their bodies obliquely into the wind.

Direction communication is not perfect, however, and both the direction indicated by dances and the direction taken by recruited bees show about the same average error, between 9° and 12°. These inaccuracies diminish for resources at greater distances, and older, more experienced dancers are more precise in the angle communicated. To compensate partly for this problem, recruits generally attend at least six or seven dances before leaving the nest and then take the sum of the angles communicated (Esch and Bastian, 1970). The combination of distance and direction errors made by dancers and recruits results in fewer than half of the recruits actually finding a resource, and an average of between two and five flights are necessary before a recruit is successful at finding a food location (Mautz, 1971; Seeley, 1983). As Seeley (1985a) has pointed out, however, the number of bees visiting a forage patch requiring 20 min per round-trip can easily increase more than tenfold within a few hours, which demonstrates the remarkable efficacy of this seemingly inefficient system.

In addition to signaling the presence of nectar and pollen, dances in the nest can indicate the location of water and possibly propolis, and dances on the face of swarms indicate the locations of potential nest sites. Park (1923c) first described dances to water sources by foragers, which can communicate distance and direction to potential recruits. Returning foragers carrying plant resins have been observed to dance, and recruits to propolis from colonies with such dancers quickly appear at resin sources (Meyer, 1954; Milum, 1955). Other aspects of propolis dancing have not been studied. As discussed in Chapter 5, the locations of potential new nest sites for swarms are indicated by dances, with a consensus to one site eventually being reached.

THE DVAV AND OTHER DANCES

There are other dances in addition to the round and waggle dances which are performed by workers, most notably the DVAV, or vibration, dance. In this dance a worker vibrates its body dorsoventrally, particularly the abdomen, frequently while grasping another worker or a queen (Milum, 1955; Allen, 1959a,b). These dances can occur hundreds of times per hour in a colony (Hammann, 1957; Fletcher, 1975) and are used to regulate two different colony activities, foraging and swarming.

The link between the vibration dance and foraging has been recognized in many studies, since the dance is frequently performed by pollen-carrying workers and waggle dancers, and vibration dance activity and flight activity have similar seasonal patterns (Istomina-Tsvetkova, 1953; Allen, 1959b; Winston and Punnett, 1982). Only recently, however, has the association between foraging and vibration dancing been clarified (Schneider, Stamps, and Gary, 1986a,b). It is now apparent that these dances are used to regulate both daily and seasonal foraging patterns according to short- and long-term fluctuations in food availability.

The regulatory activity of vibration dances functions in a number of ways. First, workers of foraging age, but not younger workers, respond to the dance by increasing their rate of movement in the nest, particularly by moving into the area where waggle dances are being performed. Second, daily peaks in the level of vibration dancing are closely related to peaks in foraging activity. Colonies which have experienced 3–4 days of successful foraging show increased vibration dances early in the morning, and smaller peaks are evident at any time of day within 30 min of increased foraging activity induced by the sudden availability of a new resource. Third, there are long-term seasonal peaks in vibration dancing associated with periods of food abundance. Therefore, the vibration dance acts as a short- and long-term primer for foraging by concentrating potential recruits in the waggle dance region of the colony, thereby increasing recruitment efficiency and regulating daily and seasonal foraging activity.

One of the most interesting aspects of this dance is that the same worker behavior has different effects in different contexts. In addition to activating foraging, the DVAV dance also regulates queen activities associated with swarming (Fletcher, 1975, 1978b,c). The number of times a mated queen is vibrated rises rapidly once queen rearing begins but drops suddenly a few hours prior to swarming (Fig. 9.6). This intensive vibration of the queen serves to inhibit her activity, possibly preventing her from destroying the new developing queens. When dancing suddenly ceases, the removal of the inhibitory effects on the queen's activity may stimulate her to exit the nest with a swarm. During the period following the issuing of this prime swarm wth the old queen, workers will vibrate developing queen cells with increasing intensity, almost continually vibrating cells with mature virgin queens inside, particularly after one queen has emerged. These dances on the cells diminish immediately prior to the exit of an afterswarm. Thus, the vibration dance also exercises some control over queen emergence and afterswarming. A drop in worker vibrations on emerged virgin queens of mating age may stimulate mating flights, although this association is not as strong as for the other queen-related vibration dance functions. In summary, dorsoventral abdominal vibrations seem to have an activating effect on foraging in worker-worker interactions and an inhibitory effect on queen activity

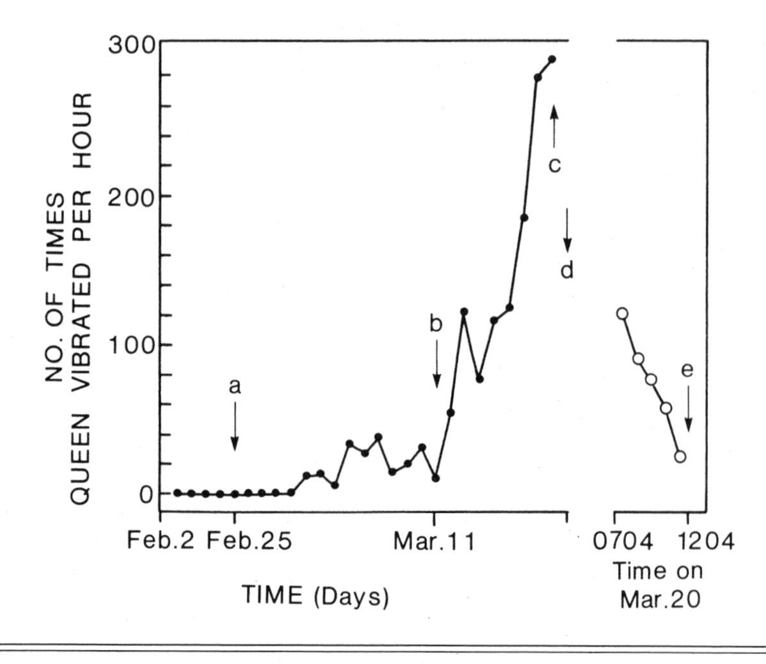

Fig. 9.6

The frequency with which a queen honey bee was vibrated during a swarming cycle. (*a*) First queen cup constructed. (*b*) First egg laid in a cup. (*c*) First queen cell sealed. (*d*) Colony swarmed on March 20. (*e*) Queen departed with swarm. (This study was conducted in South Africa; redrawn from Fletcher, 1975. Reprinted by permission of *Nature*. Copyright © 1975 by Macmillan Journals Limited.)

that may also function to prime queens gradually for swarming or mating flights when vibrating diminishes.

There are other dances in the nest for which the functions are not as well understood as the round, waggle, and DVAV dances. These dances have been reviewed by von Frisch (1967a) and include the jostling run, spasmodic dance, buzzing run, shaking dance, and trembling dance. In the jostling dance returning and successful foragers jostle their nestmates by running at them and pushing them aside; this may alert them that a dance is about to occur. The spasmodic dance involves food distribution interspersed with short tail-wagging movements and may be similar to the jostling dance in signifying forthcoming information about resources. The buzzing run is used to signal the exit of a swarm, or induce a swarm to alight. In this dance workers buzz their wings and run through the colony or swarm in a random but energetic pattern; these movements increase the activity of their nestmates and lead to flight. In the shaking dance the worker shakes her body rapidly from side to side, inducing nearby workers to groom her. Trembling dances have been observed following colony disturbance in which workers twitch, tremble, and run about on four legs; its function is totally unknown.

Navigation and Orientation

The communication mechanisms of honey bees can transmit information about the location of resources, but they do not explain how workers which have learned this information get to and from the resources or recognize them when they get there. Like many organisms, honey bees do not use only one navigational mechanism; rather they possess a complex combination of visual, olfactory, and magnetic senses which are utilized to find their way in a constantly changing environment. The links between communication and orientation have largely been forged by von Frisch, who made many key discoveries in both fields. As Dyer and Gould (1983) have pointed out, he made two conceptual discoveries which have served as guideposts for all subsequent researchers. First, he determined that bees see colors but are blind in the red spectrum and acutely sensitive to ultraviolet. This finding confirmed that the orientation mechanisms of honey bees are uniquely adapted to their needs, and that in many ways they occupy a different sensory world than humans do. Second, von Frisch realized that there are primary and secondary orientation mechanisms used by bees, and that these are organized hierarchically, so that if the primary mechanism is not operable for some reason, other mechanisms can be used for navigation.

The redundancy in navigational tools possessed by worker bees is best exemplified by visual orientation, in which the location of the sun, polarization of light waves, and landmarks can all be used in getting to and from resources. The primary mechanism is sun compass orientation, by which workers cannot only orient to the sun's position but also compensate for its diurnal movement through the sky. The key experiment demonstrating this sense was performed by von Frisch; he first trained workers to feed at a station 200 m west of a colony and marked the foragers visiting the station. In the evening the hive was moved to a new site and feeding dishes placed 200 m from the hive in all four compass directions. When the hive was reopened the next morning, most experienced foragers went to the dish in the west, and opening the hive at any time of day yielded the same result. Thus, workers are able to calculate the sun's movement through the sky and make the proper corrections to their flight direction to compensate for that factor. Subsequent experiments demonstrated that it is the sun's azimuth, or compass direction, that is used as a reference point, and its elevation above the horizon is not important. In fact, workers that continue to dance after dark can compensate for the movement of the sun after sunset (Edrich, 1981). The sun's position below the horizon is used for orientation during nocturnal foraging, which has been observed on warm moonlit nights (reviewed by Fletcher, 1978a; Dyer, 1985).

Even more remarkable is the ability of workers to use the sun compass on cloudy days. Part of this ability is due to the use of ultraviolet light waves rather than those in the visible spectrum to sense the position of the sun. Since ultraviolet waves can penetrate through overcast skies so long as the cloud cover is not too dense, workers can use these waves to locate the compass direction of the sun.

Workers can even orient under partly cloudy skies when the sun is not visible; this is accomplished by one of the principal backup mechanisms to sun compass orientation, the perception of polarized light patterns. Natural light from the sun is polarized, meaning that the direction of vibration of light waves changes in a regular pattern as the sun moves through the sky. These polarization patterns are not visible to humans but can be perceived by bees and many other organisms through a patch of blue sky even when the sun is not visible. Von Frisch demonstrated this ability by forcing workers to dance on a horizontal piece of comb in sight of the sky. Dancing workers would normally point directly at the food source to which they were trained, but when a polarizing filter was introduced between the dancers and a patch of blue sky, the dances would shift according to the shift in the polarized light pattern induced by the filter. Similar techniques have since verified that workers determine the relationship between polarized light and the sun's position during their early flights on sunny days, and that it is the ultraviolet component of polarized light which is perceived by workers. The hierarchical importance of actually seeing the sun as opposed to seeing only polarized light through a patch of blue sky was revealed by giving bees two conflicting bits of information, sight of the sun and incompatible information about the direction of light polarization using a filter. In these experiments workers generally chose the direction indicated by the sun, showing that sun compass orientation is the primary orientation mechanism, and polarization is a backup mechanism.

Workers evidently use other backup mechanisms for navigation, since recruitment to feeding stations can be intense under heavily overcast and even rainy conditions, when neither the sun's compass position nor the light polarization pattern is visible to bees. One such mechanism is orientation by landmarks. In an early series of experiments von Frisch and Lindauer located feeding stations along prominent landmarks, such as a line of trees or a road extending in one direction. Once workers were trained to these stations, hives were moved to new sites, some with similar landmarks, some not. Sites with landmarks were in a different compass direction than the original training site. Trained foragers generally went in the proper compass direction at the new site without landmarks, but at the landmark sites some foragers flew in the proper compass direction, and other, possibly more experienced, foragers followed the landmarks in the wrong direction. The conclusion drawn from these experiments was that although landmarks are not necessary for orientation, they are used by experienced foragers in addition to and sometimes in lieu of solar orientation.

Subsequent experiments (reviewed by Dyer and Gould, 1983) showed that those foragers which chose to follow the landmarks would return to the colony and dance to the new sun compass direction rather than the one to which they were previously trained. Moreover, when hives were relocated on cloudy days, almost all foragers would visit the site indicated by landmarks rather than the sun compass station. The fascinating aspect of these visits was that upon returning to the hive, these workers would dance as if they had been flying to the feeding station at the previous site. And these misdirected dances continued until later in the day when the sun came out again. Thus, workers can use landmarks to memorize the course of the sun, and, although they will follow the landmark direction in the absence of the sun, the primary importance of sun compass orientation is evident, since the foragers correct their dances as soon as the sun becomes visible again.

Besides vision there are other sensory systems involved in distance orientation which allow workers to find and return from foraging sites in the absence of landmarks and solar cues, among them orientation to the earth's magnetic field. There are four lines of evidence for magnetic field orientation in bees (reviewed by Gould, Kirschvink, and Deffeyes, 1978; Gould, 1980). First, minor but regular errors in the direction indicated by dances depend partly on the orientation of the dance with respect to the earth's magnetic field; these errors can be eliminated by canceling the field around a colony. Second, workers forced to dance only on horizontal comb eventually adjust their dances so that they are oriented relative to the magnetic field rather than to gravity; canceling the field also eliminates this reorientation. Third, swarms placed in cylindrical hives and deprived of all cues except the earth's magnetic field often build their comb in the same magnetic direction as their parent hive. Finally, circadian rhythms can be set according to regular daily variations in the earth's magnetic field in the absence of any other cues.

The location of the sensory cells responsible for this magnetic sense are not known, but two types of potential magnetic detectors have now been found: first, a region of transversely oriented magnetic material in the front of the abdomen (Gould, Kirschvink, and Deffeyes, 1978), and second, bands of cells around each abdominal segment that hold iron-containing granules (Kuterbach et al., 1982). Either or both of these regions may be capable of detecting subtle changes in a worker's position relative to the field by the torque induced by the field when the worker shifts position.

Locale odor, which may attract honey bee workers over long distances, has been proposed as both a supplementary cue and an alternative to dance language. The suggestion that dance language may not communicate distance and direction at all has been made by Wenner and colleagues, who believe that odor alone can explain recruitment (see particularly Johnson, 1967; Wenner and Johnson, 1967; Wenner, Wells, and Rohlf, 1967; Wenner, 1971; and rebuttals summarized by von Frisch, 1967b, and Gould, 1976). Although the experiments of von Frisch and colleagues and additional experiments with more extensive controls have overwhelmingly confirmed the efficacy of the dance language, it is also clear that locale odors are learned by workers and used as an additional cue for orientation. These odors might include floral odors or local environmental odors such as hedges, turned soil, and leaf litter, and would of course be more important if the resource was located upwind.

Workers have short-distance orientation capabilities which are used both at the resource location and in finding the nest following flights. As we have seen, recruits following the bee dances obtain information about the odor of the flowers being visited by antennating the dancer, and this information aids in finding the flower which is producing nectar and pollen. On homeward flights workers can orient to both general orientation pheromones, such as the Nasonov scents, as well as to nest-specific odors, which allow them to discriminate between their colony and foreign colonies.

The visual system is also utilized for local orientation to flowers or nests and in this capacity exhibits a remarkable array of perceptive capabilities. Worker bees are particularly good at seeing colors, patterns, and movements, all characteristics especially useful in the context of what a worker needs to recognize. For color, bees possess trichromatic vision roughly comparable to humans', but with some important differences (Fig. 9.7). Bees are most sensitive to the short-wavelength end of the spectrum, particularly to ultraviolet light. Their sensitivity to other colors decreases from ultraviolet in the order of blue-violet, green, yellow, blue-green, and orange; bees are insensitive to red and can see a color called "bee purple," which results from a combination of the two colors at the ends of the bee's visual spectrum, ultraviolet and yellow. As for humans, colors across from each other in a color wheel complement each other to form white.

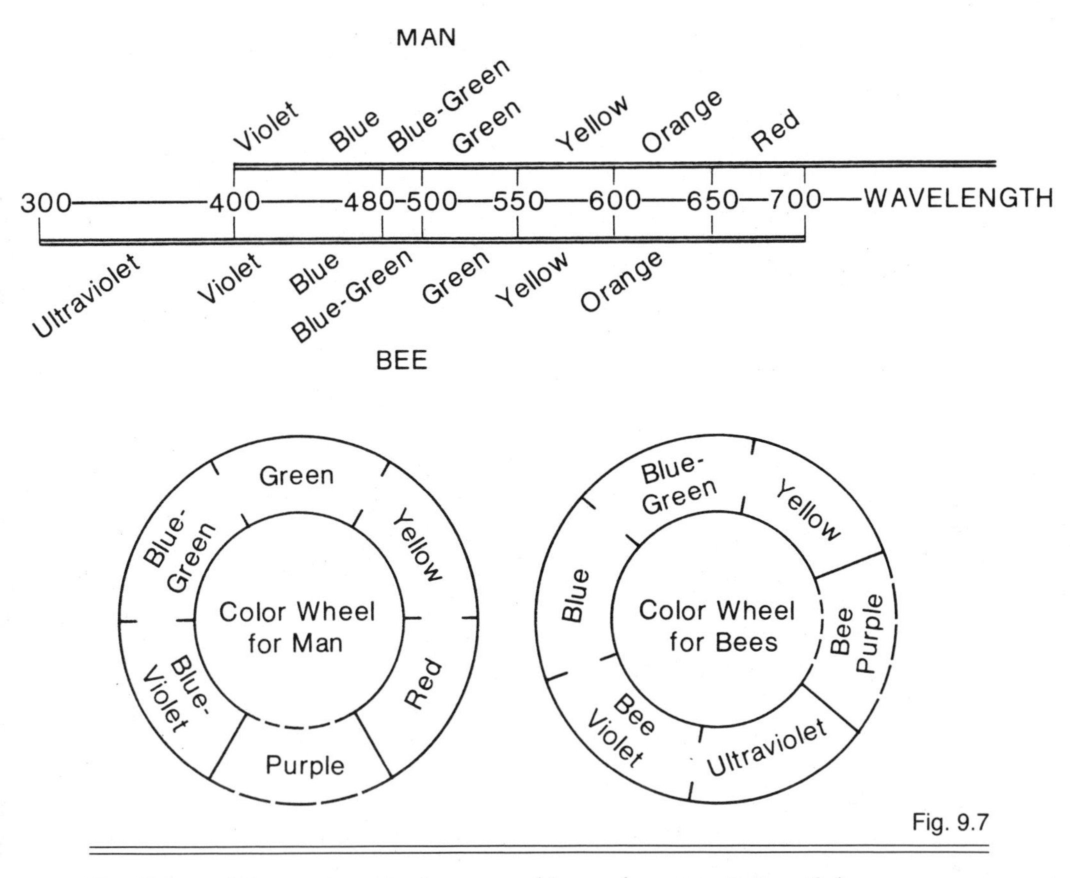

Top: Colors of the spectrum for human and honey bee eyes. *Bottom:* Color wheels for humans and honey bees. (Redrawn from von Frisch, 1967a, and Michener, 1974.)

Two types of evidence have confirmed color vision in bees: (1) when workers have been trained to visit colors, they do so in the order described above; and (2) electrophysiological techniques used to determine sensitivity curves for single receptor cells have yielded identical results. Similar techniques have been used to demonstrate that a bee approaching a flower has a very different vision from what humans perceive. It appears, in fact, that flowers have evolved their visual characteristics so as to be most visible to their bee pollinators. For example, foraging bees often home in on the ultraviolet patterns produced by flowers, and these patterns are often in the configuration of "nectar guides," or regions of the flower which do not reflect ultraviolet and which indicate the location of nectaries, stamens, and pistils of flowers (Free, 1970). Background green foliage appears gray to bees, so that floral colors are that much more prominent.

Bees are also able to perceive form, especially the degree of disruption caused by the frequency with which edges cross their field of vision (see particularly research by Hertz as cited and reviewed by Ribbands, 1953; von Frisch, 1967a; and Free, 1970). Wide angles are also more distinguishable than narrow angles, and closed shapes such as circles, triangles, and squares are not easily distinguishable from each other, although they can be distinguished from open shapes with wide angles. This type of visual acuity is common among flying insects, which can best see disruptive, wide angles while moving in flight. In contrast, homeward-bound workers prefer dark figures with little structure or contour, a pattern similar to the dark nest entrance to which they are returning.

The other aspect of vision which is particularly acute in bees is perception of movement. By moving patterns of black and white stripes across the eyes of bees, it has been determined that workers can detect rates of up to 300 stripes/sec (Autrum and Stoeker, 1950). This flicker fusion frequency is notably higher than the human rate of only 15–20/sec. Again, such a high acuity would be expected in flying insects which must quickly respond to relatively stationary patterns formed by vegetation or nest entrances.

Thus, bees combine visual acuity for the perception of polarized light, color, patterns, and movement with olfactory and magnetic senses to obtain information from the environment important for navigation and orientation. These senses are well integrated with communication systems, which enable colonies to discover and exploit resources quickly—one of the principal advantages of sociality.

10

The Collection of Food

To bees, food means nectar and pollen; all of the nutritional requirements of brood and adults are met by these two plant-produced substances. Like most aspects of colony functioning, food collection is organized hierarchically by integrating simple worker behaviors with colony requirements. But there are a staggering number of behaviors at the individual and colony level which must be coordinated for efficient nectar and pollen collection. To name only a few individual behaviors: the age at which foraging commences, the flowers visited, the substances collected (nectar, pollen, or both), and the method for working the flower to maximize the resources collected with minimal energy. At the colony level, foraging can be influenced by the proportion of scouts to recruits, information transfer within the nest, allocation of foragers to nectar and pollen collection, and time at which workers switch from a poor to a better resource. These complex behaviors can, however, be approached by asking three simple questions which encompass most aspects of foraging behavior: where do workers go on foraging trips, how do individual workers organize their nectar and pollen collection, and how do colony-level behaviors control foraging?

Where Workers Go

A worker leaving the nest on a foraging trip can face an overwhelming array of flowers to choose from, some of more value than others. The quantity and quality of nectar and/or pollen produced by useful flowers can vary tremendously, both between plant species and within patches of the same species. In some regions plants can produce nectar so copiously that colonies can collect 5 kg or more of nectar daily and then produce a harvestable honey surplus of more than 200 kg annually, whereas in other regions colonies must be fed sugar syrup for much of the year to survive. Although colonies are able to respond with great flexibility to such variable conditions, the food collection potential of colonies is ultimately limited by the nectar and pollen production of plants.

Beekeepers refer to the periods of heaviest nectar production as honeyflows, and most regions have somewhat predictable seasons during which the best nectar-producing flowers are in bloom (Fig. 10.1). These flows can be deceptive, however, since annual fluctuations due to weather conditions or changing crop patterns can result in two- to three-fold changes in honey production from one year to the next. Sudden habitat changes can also result in dramatic differences in a region's nectar secretion; for example, a forest fire may mean that a high nectar-producing plant such as fireweed will grow copiously in burned-over areas, resulting in a few years of tremendous honey production before the next post-fire successional stage develops. Also, the success of both feral and managed colonies near contemporary agricultural systems depends largely on what crops are planted; endless acres of clover, oilseed crops, alfalfa, and fruit trees can provide impressive but artificial honeyflows.

There are numerous plants worldwide which are known for their nectar production. A list of the best plants, which can produce over 500 kg/hectare of honey, includes the common maple, milkweed, phacelia, sage, thyme, acacia, and figwort (Crane, 1975; Robinson and Oertel, 1975). These range from trees to bushes to ground cover plants and are taxonomically diverse. All of the good honey-producing plants are characterized by well-developed floral and sometimes extrafloral nectaries which can concentrate and secrete sugars, and their flowers are generally designed so that bees are easily attracted and the nectar is accessible. But even the best sources vary widely in honey production in different years and regions, depending on such factors as air temperature, humidity, groundwater, precipitation, and soil fertility (Shuel, 1975).

Many plants invest energy in nectar production to attract bees, which serve to transfer pollen from one plant to another. The bees, of course, use the pollen for food, but in the process of collecting nectar and pollen sufficient pollen grains are transferred between flowers to effect pollination. The pollen is produced in the stamen of the plant and, as for nectar, the quality and amount of pollen produced can vary tremendously within and between plant species. Some plants actually produce little or no nectar but are nevertheless attractive to bees because of their pollen production. Workers choose which pollens to collect not by their nutritive value, age, moisture content, or color, but on the basis of the odor and physical configuration of the pollen grains (reviewed by Stanley and Linskens, 1974, and Jay, 1986).

In addition to deciding which species of flower to visit on a forag-

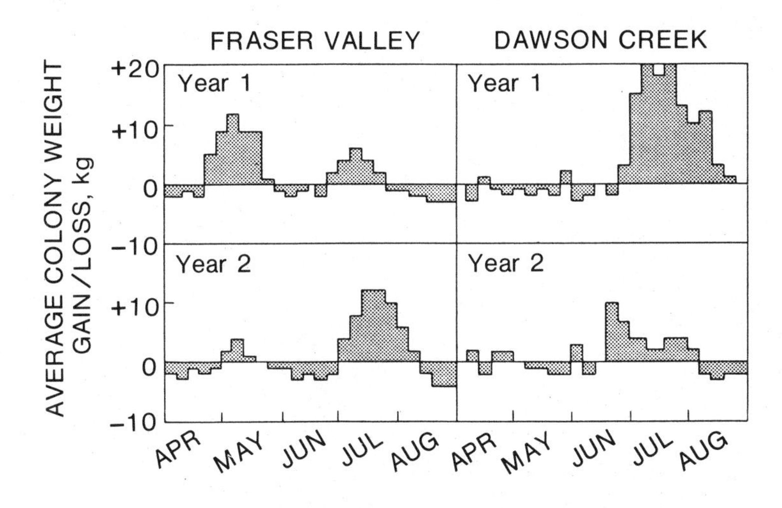

Fig. 10.1

Honeyflows in two regions of British Columbia, Canada, during two different years, as measured by average weekly weight gain or loss of colonies. The Fraser Valley is in the southwest and normally has a flow in the spring and again in the summer. Dawson Creek is in the northern part of the province and has one strong flow in the summer. Colonies gain weight when more nectar is being collected than is being used by the bees and lose weight when nectar collection cannot keep pace with consumption. (Data condensed from B.C. Ministry of Agriculture and Foods records of colonies maintained on scales for a number of years.)

ing trip, workers must also determine where to search and how far to forage routinely. In agricultural areas the median foraging radius of honey bee colonies is only a few hundred meters (reviewed by Visscher and Seeley, 1982), although significant forager populations have been found at 3700 m from apiaries (Gary, Witherell, and Marston, 1972), and recruitment to feeding stations can be induced at up to 10,000 m if no other competing food sources are available (Knaffl, 1953). The median foraging radius in forested regions is larger, 1.7 km, with most foraging occurring within 6 km of colonies. Bees from adjacent colonies tend to forage in slightly different sections of fields, but it is not clear how foraging sites might be apportioned between colonies (Levin and Glowska-Konopacka, 1963).

The Organization of Foraging Trips

Although foragers may have gained considerable information about the location and nature of the resource from following dances, there can still be considerable flexibility in the number of flowers they visit, whether they collect nectar, pollen, or both, how they work a flower, and whether they return to a particular site, among other factors. One of the most important choices for a forager is what to collect. As with most foraging characteristics, this is partly determined by the availability of flowers in the field; nevertheless, there are some consistent trends in the proportion of workers which are nectar and pollen foragers. Workers tend to prefer nectar collection. In two studies about 58% of foragers collected only nectar, 25% only pollen, and 17% both nectar and pollen (Parker, 1926; Free, 1960b). Workers tend to "specialize" on one type of foraging task at a time and often exhibit constancy for nectar or pollen collection during many consecutive trips. Workers also tend to visit only one species of flower on a trip, and they continue to visit that flower for prolonged periods until it ceases to produce nectar or pollen or until a superior source becomes available. The number of plant species visited on a single trip can be most easily determined for pollen foragers, since the pollen loads can be examined at the entrance and the species of pollen classified. Mixed loads are most commonly found in less than 3% of the returning foragers, although multi-species visits have been found in up to 13% (Betts, 1935; Maurizio, 1953; Schwan and Martinovs, 1954; Tushmalova, 1958; Free, 1963).

There are some similarities in foraging statistics for nectar- and pollen-collecting workers, as well as some important differences (Table 10.1). The most obvious similarity is that both types of foragers exhibit enormous variation in the number of flowers visited per trip, the number of trips per day, the time needed to collect a load, and the weight of a full load. Both nectar and pollen collectors need anywhere from 1 to 500 flowers to gather a full load, depending on the extent of nectar or pollen production, although some workers may visit way over 1000 flowers to collect a nectar load. Both nectar and pollen foragers make an average of 10 to 15 trips a day, although nectar foragers may make up to 150 trips. The time per load tends to be lower for pollen collectors, generally about 10 min (but up to 187 min) as opposed to 30–80 min (but up to 150 min) for a nectar trip. Finally, pollen loads are usually 10–30 mg in weight, whereas nectar loads tend to weigh 25–40 mg. Thus, workers are returning to the colony more heavily laden with nectar than with pollen relative to their body weight.

Nectar collection may be energetically more efficient than pollen collection; the energy returns for pollen show approximately an 8:1 ratio on flight energy expended per calorie of pollen collected, but a 10:1 ratio for nectar collection (Seeley, 1985a). However, workers will travel further for a pollen load than for nectar, possibly because a full load of pollen is lighter in weight and takes less time per flower than

Table 10.1 The average or range of foraging statistics for nectar- and pollen-collecting workers

	Pollen			Nectar		
Characteristic	Statistic	Flower (if known)	Reference	Statistic	Flower (if known)	Reference
No. flowers for full load	84	Pear	Vansell, 1942	1400	*Limnanthes*	Ribbands, 1949
	100	Dandelion	Vansell, 1942	1	Eucalyptus	Michener, 1974
	8–32	Dandelion	Ribbands, 1949	1	Tulip poplar	Miller, 1902
	1	Poppy	Ribbands, 1949	50–1000	—	Miller, 1902
	66–178	Nasturtium	Ribbands, 1949			
	494	Clover	Weaver, Alex, and Thomas, 1953			
No. trips per day	6–8	—	Park, 1922	7–13	—	Ribbands, 1949
	47	Poppy	Ribbands, 1949	10	—	Lundie, 1925
	8 (20 max)	Maize	Park, 1928b	10	—	Heberle, 1914
	10	—	Singh, 1950	150 (max)	—	Ribbands, 1953
				3–10	—	Singh, 1950
				13	Clover	Park, 1928b
				7	Clover	Park, 1928b
Time per load (min)	6–10	—	Park, 1922	27–49	Clover	Park, 1928b
	187	—	Singh, 1950	106–150	*Limnanthes*	Ribbands, 1949
	13	Clover	Weaver, Alex, and Thomas, 1953	40	Lime	Ribbands, 1951
				80	Goldenrod	Singh, 1950
Weight of load (mg)	11	Elm	Park, 1922	40 (70 max)	—	Parker, 1926
	29	Maple	Park, 1922	25	—	Lundie, 1925
	8–21	—	Maurizio, 1953	33	—	Combs, 1972
	11	—	Gillette, 1897	27	—	Fukuda, Moriya, and Sekiguchi, 1969
	9	Asparagus	Parker, 1926			
	20	Ragweed	Parker, 1926			
	15	—	Parker, 1926	12	—	Fukuda, Moriya, and Sekiguchi, 1969

a nectar load. Also, colonies do not store large reserves of pollen compared to honey, and the necessity to replenish pollen more regularly than honey may stimulate workers to travel further for pollen collection. In one experiment foraging bees traveled a mean of 1663 m to visit pollen-yielding carrots, but only 557 m to visit nectar-yielding onions (Gary, Witherell, and Marston, 1972). Workers have been known to travel relatively long distances to visit high-yielding pollen plants such as safflower.

Foragers show great versatility in their methods of working flowers, depending on the location of the nectar and pollen. Parker (1926) classified the flowers which honey bees visit into the following types according to how they are worked for pollen.

Open flowers. The worker bites the anthers with her mandibles and uses the forelegs to pull them toward her body.

Tubular flowers. Workers insert the proboscis into the corolla searching for nectar, and pollen is collected incidentally when it adheres to the mouthparts or forelegs.

Closed flowers. The bee forces the petals apart with her forelegs and then gathers pollen on the mouthparts and forelegs.

Spike or catkin flowers. The bee runs along the spikes or catkins, shaking off pollen onto her body hairs.

Presentation flowers. The pollen is collected by workers pressing their abdomens against the inflorescence, causing a pollen mass to be pushed out of the flowers.

Nectar collectors generally show less variability in how they work flowers. They alight on the flower, insert their proboscis into the corolla, and pump the nectar up into the mouth. Unusual types of nectar collection which bypass the corolla are sometimes performed by workers, and in these cases pollen is not collected or transferred between flowers. For example, bumble bees sometimes chew holes in the side of blueberry flowers to get at the nectary, and honey bees will use these holes rather than the corolla to imbibe nectar. The Red Delicious variety of apples is frequently "sideworked" by honey bees which hang on to the side of the flower and insert their tongues down the side of the flower without contacting the pollen-producing stigmatic surface. In alfalfa the mouthparts of workers can get temporarily trapped when the flower "trips" upon probing, and experienced workers learn either to visit flowers which have already been tripped or to probe flowers from the side, again eliminating pollen collection and transfer.

Workers also tend to visit the same locale at the same time of day for many consecutive trips or days, as long as the floral rewards continue. A worker may return to locations only a few meters in diameter in homogeneous environments such as apple orchards or alfalfa fields (Free, 1966; Levin, 1966) and can continue to visit the same small area for most of her foraging life. Many flowers secrete nectar or pollen only at certain times of day, and workers visiting these flowers will reappear at the same time on consecutive days expecting a "reward." The same individual can be trained to visit three different sites at three different times of day, and, if the concentration of sugar syrup is varied throughout the day, workers will visit in the greatest numbers when the sugar concentration is highest (reviewed by Ribbands, 1953, and Michener, 1974).

Individual foraging behaviors involved in the collection of nectar and pollen are influenced by both genetic and environmental factors. The genetic basis for the collection and storage of nectar and pollen has been shown by demonstrating variability in these factors between colonies and then successfully selecting for high and low hoarding characteristics. The first demonstration that selection could influence pollen collection was done using preferences for the collection of alfalfa pollen (Mackensen and Nye, 1966, 1969; Nye and Mackensen, 1968, 1970). In these experiments two-way selection over six generations resulted in colonies with high and low preferences for alfalfa pollen, with 86% of workers in the high category visiting alfalfa as opposed to only 8% in the low category. The tendency of colonies to hoard pollen can also be modified by selection. In one experiment high and low levels of pollen hoarding were selected in only four generations, and colonies in the high line stored from 2 to 13 times as much pollen as the low line colonies (Hellmich, Kulincevic, and Rothenbuhler, 1985). Both the alfalfa preference and pollen-hoarding traits were shown to be polygenic and have high heritabilities. Similar results were found when fast and slow hoarding of sugar syrup was investigated, with fast hoarders showing significantly more weight gain in laboratory cages and in field studies than slow hoarding colonies (Rothenbuhler, Kulincevic, and Thompson, 1979).

Weather and other environmental factors also influence nectar and pollen foraging (Ribbands, 1953, and references cited therein). Although workers can fly during winter at temperatures only slightly above 0°C, and pollen collecting has been seen at temperatures as low as 5°C, foraging for either nectar or pollen generally does not begin until temperatures reach 12–14°C. For light intensity, global solar radiation values of 0.66 langleys or greater are positively correlated with flight departures, and at lower values this relationship is negative (Burrill and Dietz, 1981). Relationships between wind and rain, and foraging intensity, show decreasing foraging resulting from higher winds and heavier rainfall. Time of day can influence flight patterns, with fewer bees flying in the early afternoon. The cause of this relative flight inactivity is not clear and may simply reflect less nectar production by flowers at that time of day. Many of these factors interact with each other and with colony and flower characteristics to determine foraging patterns. For example, Scott (1986) found various combinations of positive and negative correlations between temperature, global solar radiation, and relative humidity and forager entrance and exit patterns for colonies located in cherry, pear, and apple orchards. These variable results were probably caused in part by differences in nectar and pollen availability in the crops, so that workers would forage under more adverse conditions when the reward was greatest.

Other aspects of individual behavior which influence foraging efficiency involve the learning of orientation cues to flowers and patterns of interfloral movement. One reason that foragers tend to visit only one species of flower per trip, and often for days, is that experience with flowers reduces both the search time and the time spent working the flower. Learning and memory have a key role in this pattern of floral specialization, and workers learn quickly to orient to the shape, odor, and color of a flower they are working (reviewed by Menzel, Erber, and Masuhr, 1973). It takes very few visits to learn these cues; a single experience with a particular odor associated with a sugar reward will induce constancy to that odor in 90% of workers, and five exposures to a food reward associated with a particular color will induce the same level of constancy. Since workers visit many flowers during each foraging trip, the visits during a single trip provide sufficient learning experience for subsequent trips. The memory of orientation cues to flowers routinely persists for periods of weeks, and trained workers will return to the same locale searching for food following almost 175 days of confinement in the nest during the winter.

Even within flower patches workers adjust their patterns of movement to increase the likelihood of encountering flowers yielding high rewards. Within a profitable patch workers change direction often and move short distances as they fly between flowers; this movement results in the foragers remaining in a relatively small area. When the flowers in a patch are not providing as great a reward, however, workers change direction less frequently and fly longer distances between flowers; these longer, straighter flights tend to move bees away from unrewarding patches and into new patches which might provide better rewards (Waddington, 1980). However, the nectar reward itself is not the only cue used to determine the degree of directional change; foragers also change direction between flowers less often with an increase in the amount of time spent on a flower (Schmid-Hempel, 1984).

Colony-Level Control of Foraging

The importance of colony-level factors in coordinating worker foraging patterns can best be appreciated by examining the annual consumption of nectar and pollen by a single colony, and the amount of work required to collect those foods. Studies of food consumption have only been conducted for temperate-evolved bee races, and some variability in the results of different studies is expected due to the widely divergent environmental conditions in different areas. Nevertheless, even the most conservative data are impressive in the amounts of food which colonies must collect to survive. Colonies require between 15 and 30 kg of pollen a year, although they may collect up to 55 kg (Eckert, 1942; Hirschfelder, 1951; Louveaux, 1958; Seeley, 1985a). Of honey, colonies require from 60 to 80 kg annually (Weipple, 1928; Rosov, 1944; Seeley, 1985a). Since average loads of nectar and pollen are each in the range of 10–40 mg (Table 10.1), it clearly requires a considerable number of trips to collect enough of both to meet colony requirements. Using the minimum requirements cited above and average load values of 15 mg for pollen loads and 16 mg of sugar in an average load of 32 mg of nectar (derived from data in Table 10.1), workers must make one million trips annually to collect pollen and almost four million trips to collect nectar. Furthermore, in temperate areas these trips must be made within the relatively short period when flowers are blooming, and even then only during the hours and days when the weather permits food collection. Thus, although the workers in strong colonies can make up to 163,000 trips daily under ideal conditions (Gary, 1967), most feral colonies barely collect enough resources to survive, and many colonies starve to death every winter (Seeley, 1978; Lee and Winston, 1985a; Seeley and Visscher, 1985).

Given the importance of food collection, it is not surprising that honey bees have evolved numerous mechanisms for increasing colony-level foraging efficiency. These mechanisms can be divided into two major groups: those for obtaining information about colony requirements and using that information to stimulate workers to forage for nectar and pollen; and those for allocating foraging tasks to maximize gain, recruiting to resources, transferring information, and making decisions about when to switch resources.

Because workers have many sources of information from which to learn about colony needs, foraging patterns can be adjusted according to colony requirements. This is particularly true for the stimulation of pollen collection required for brood rearing. There is a positive relationship between the amount of eggs and larvae in colonies and both the proportion and number of pollen foragers (Filmer, 1932; Fukuda, 1960; Free, 1967b; Todd and Reed, 1970; Al-Tikrity et al., 1972). The proportion of workers collecting pollen also increases with higher egg-laying rates by the queen (Cale, 1968). The stimulatory effect of the queen and brood is probably mediated by pheromones (see Chapter 8), but works quickly whatever the sensory mechanism involved; doubling the amount of brood in a colony can increase pollen collection as much as three times within 24 hr. Workers evidently can assess the amount of stored pollen relative to brood-rearing requirements, since adding pollen or artificial pollen supplement to a colony decreases pollen collection (Free, 1967b; Barker, 1971; Moeller, 1972). Thus, workers are able to vary rapidly their foraging patterns in response to changes in colony pollen requirements. The factors which influence nectar collection are not as well known, although queen odors, the presence of worker larvae, and empty comb have been found to stimulate nectar collection (Jaycox, 1970a,b; Rinderer, 1981).

While assessing a colony's food requirements is of obvious importance, efficient discovery and exploitation of resources is also fundamental to colony survival. Single workers could, of course, search for flowers and collect nectar and pollen without communicating with other workers, but this type of foraging would be highly inefficient compared to the system of communication and recruitment to resources which honey bees actually use. It is an efficient system, whereby one group of foragers specializes in finding resources, and another, larger group is available to be recruited.

The division of labor between scouts and recruits in foraging has been recently investigated by Seeley (1983). He defined a scout bee as one which returned to the colony with nectar and/or pollen without following any dances, and determined that between 5 and 35% of the foragers in colonies were performing this scout duty at any one time. These data agree with similar determinations made earlier by Oettingen-Spielberg (1949) and Lindauer (1952). Scouts tend to be more experienced foragers, and the variability in the proportion of scouts at any time depends on the availability of nectar and pollen in the field. During dearth periods the proportion of scouts can reach a maximum of 35% of the foragers, thereby increasing the probability of finding a new resource. In contrast, a greater proportion of foragers visit known locations when good forage is abundant, and fewer foragers search for new food sources.

Seeley has extended these findings to develop what he has called the "information center" strategy of foraging (Visscher and Seeley, 1982; Seeley, 1985a,b). This concept evolved from studies of honey bee foraging patterns in forested areas more similar to the environment bees evolved in than the agricultural systems usually used to

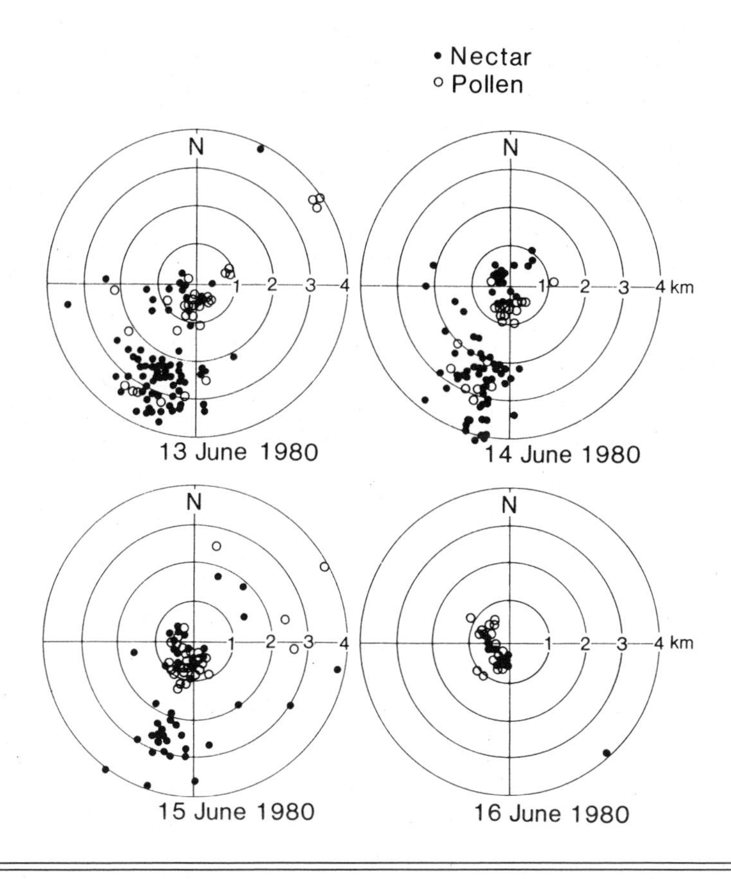

Fig. 10.2

Maps of a honey bee colony's daily foraging sites, as inferred from reading recruitment dances in glass-walled observation hives. (Redrawn from Seeley, 1985a, *Honeybee Ecology.* Copyright © 1985 by Princeton University Press. Figure 7.3 adapted with permission of Princeton University Press. See also Visscher and Seeley, 1982.)

study bees. The foraging patterns of colonies put in this natural situation are characterized by daily changes in the number of workers visiting particular patches, and relatively few patches being visited on any one day (Fig. 10.2). Moreover, colonies use scouts to monitor the forage availability over a tremendous area covering at least a radius of 4–6 km from the nest. Thus, foraging is organized so that workers concentrate on a few patches which have been discovered by scouts and found to be the best, and foraging is adjusted daily as old patches give out and new, better food sources are discovered.

These daily changes in foraging patterns are based on the information brought back to the nest by scouts and recruits, and the colony is used as an information exchange center so that workers can concentrate their foraging on the highest quality and closest resources. Potential recruits can learn the expected gain from flower patches by exchanging food and following the vigor of dances by experienced foragers, and the cost involved in energy to travel to and work the flowers can be assessed from the dances as well. By comparing information concerning all of the patches discovered by scouts, recruits can determine which patches are the most profitable of those available and visit those until their profitability declines or a better food source is discovered.

One of the interesting aspects of honey bee foraging which emphasizes the importance of the colony over the individual workers is that workers do not always maximize their energetic efficiency on a foraging flight. If individual energy expenditure were being maximized, a nectar forager would return to the nest as fully loaded as possible, but in fact many workers return from foraging trips with their honey stomachs only partially loaded. There are two explanations for this seemingly maladaptive behavior. First, foraging workers can only fly a limited distance in their lifetimes since metabolic functions associated with flight degenerate over distance (Neukirch, 1982). Thus, workers might apportion their energy expenditures for foraging in such a way as to maximize the ratio of the weight of their nectar load to their flight distance, and this maxima is reached when the crop is only partly filled (Schmid-Hempel, Kacelnik, and Houston, 1985). Second, imbibing a full load results in a loss of information about alternate food sites, since it takes more time to collect than a partial load, thereby decreasing time spent in the colony. Thus, partial crop loading may be an adaptation to increase the information center functions of the nest (Nuñez, 1982).

In summary, colonies use a variety of strategies to integrate individual worker activities to meet colony requirements most efficiently. The transfer of information within the nest is of critical importance, for workers to determine both what their colony needs and where to go to get it. Workers can respond quickly to changes in colony requirements for nectar and pollen and to the changing availability and profitability of resources outside the nest. Nevertheless, although the honey bee foraging system may be as well designed as possible, one of the more puzzling aspects of this system is that many colonies in temperate climates do not survive the winter; most starve to death. But this high colony mortality may have less to do with any innate failure of the foraging system than with swarming, the curious method of honey bee reproduction, which is discussed in the next chapter.

11

Reproduction: Swarming and Supersedure

There is perhaps no more spectacular event in the life of a honey bee colony than reproduction by swarming. In this type of colony division a majority of the workers and the old or a new queen leave the nest to search for a new home. When a swarm issues, the air is filled with the buzz of thousands of flying bees searching for their queen and a place to cluster. The workers rapidly organize themselves, creating order from what appears to be total chaos, and quickly get on with the business of finding a new nest. The actual swarm is but one brief phase in an extensive process which has gone on in the colony for many months and continues for weeks after the first swarm has left. This form of female reproduction is unusual for bees, which more commonly reproduce by rearing individual females that initiate new nests on their own. The great advantage of swarming is that the primary reproductive individuals, the queens, get assistance from the workers in building the new nest and starting brood rearing and foraging. For this advantage, however, bees must invest the tremendous time and energy swarming requires.

There has been considerable research into the events leading up to swarming, largely because the loss of swarms from managed colonies severely depresses honey production. Recently, swarming has been intensively examined by behavioral ecologists, who have found the bee colony a model system for dissecting the complex factors which determine life history and reproductive characteristics of highly social insects.

The Natural History of Swarming in Temperate Climates

Preparations for swarming by honey bee colonies in cold temperate climates actually begin in the dead of winter, when colonies begin rearing their first worker brood (Jeffree, 1956). This early brood rearing is fueled by the vast quantities of honey and pollen stored in the nest during the previous summer, and it compensates for the gradual decline in adult population during the winter. Colonies reach their minimum worker populations sometime in March, but increased brood rearing throughout the late winter and early spring gradually compensates for these winter losses, and by April the worker population has begun a dramatic rise which climaxes in swarming (Nolan, 1925; Jeffree, 1955; Allen and Jeffree, 1956) (Fig. 11.1).

Most swarming takes place in mid-spring, usually sometime in May or early June (Jeffree, 1951; Simpson, 1959; Burgett and Morse, 1974; Fell et al., 1977; Caron, 1980; Winston, 1980) (see Fig. 11.1). But swarms can issue as early as the first week in April, and a small secondary peak in swarming occurs in most areas in August and early September. The timing of swarming varies somewhat from season to season and between regions, and years or locations with late springs show later peaks in swarming. Almost all feral colonies swarm sometime in the spring, and those colonies rarely swarm again that season (Seeley, 1978; Winston, 1980). Up to 40% of the swarms which successfully establish new nests will swarm again, causing the secondary peak in swarming at the end of the summer. This is a curious behavior, since it is highly unlikely that such swarms and the new colonies they issue from will survive the winter.

The direct preparations for swarming begin about 2–4 weeks before the first swarm issues, during a period when colonies are becoming congested because of the worker population's rapid growth. For a reproductive swarm to issue, colonies must have produced or be capable of producing one or more new queens, so swarming preparations primarily involve the initiation of queen rearing. Once developing queens are present in the colony, the swarm can issue.

The first sign of swarm preparations is the appearance of new cups in which queens can be reared. Workers build and tear down queen cups throughout the season, but the number of unoccupied cups increases during the spring, prior to queen rearing. At any one time during the summer 10–20 empty cups can be found, although up to 55 have been recorded (Simpson, 1959; Allen, 1965a; Caron, 1979). Workers tend to build cups along the edges and bottoms of combs, but they can be found on the comb faces and along any irregular comb surfaces. Colonies may build cups without swarming and continue to construct cups even after swarming, but they do not swarm until queen rearing is initiated.

Queen rearing begins when eggs are laid or placed into the queen cups. Most of the eggs in queen cups are laid by the queen, but workers can and do move a small number of fertilized eggs or very young larvae from worker cells into cups (Butler, 1957a; Winston, 1979b; Punnett and Winston, 1983). Once the eggs hatch into larvae, the workers provide the special feeding which turns them into queens. The cells are elongated downward, as the immature queens develop,

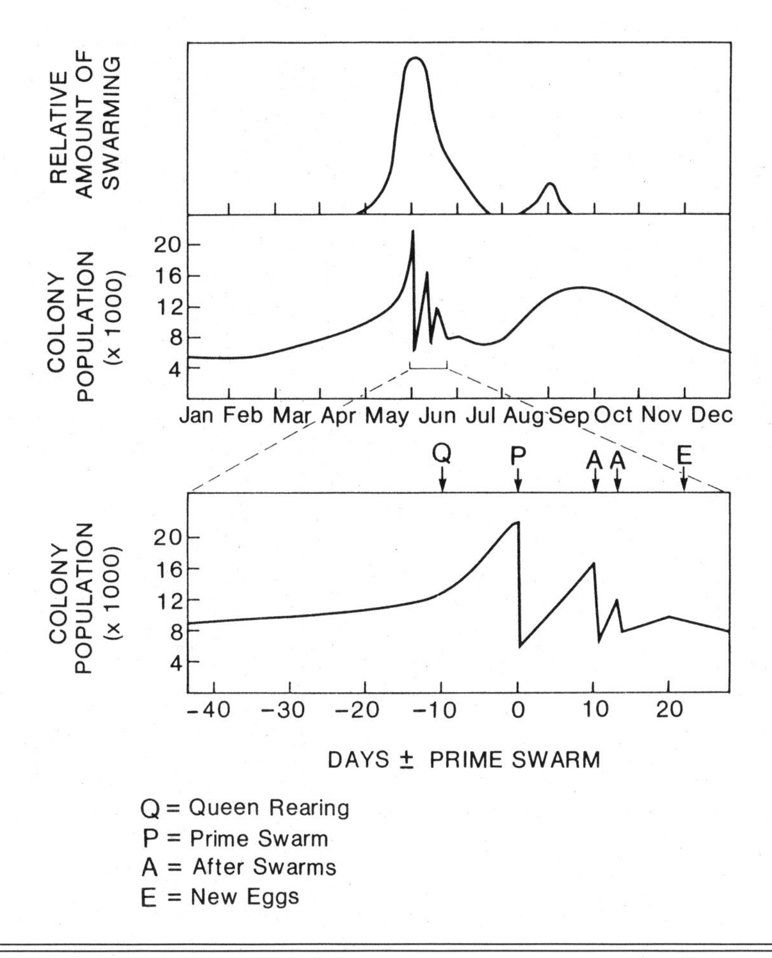

Fig. 11.1

Top: The relative amount of swarming and typical colony populations during one year in an area with a cold temperate climate. *Bottom:* The timing of queen rearing, swarming, and the resumption of brood rearing relative to the day a prime swarm issues in a typical colony. (For references, see text.)

until they are sealed at the end of the larval feeding period. Queen rearing does not proceed smoothly, however, and the larvae and/or pupae in the cells are frequently destroyed during their development by the workers or the queen (Allen, 1956; Gary and Morse, 1962; Allen, 1965a; Otis, 1980; Winston, 1980). Usually, other queens are being reared at the time of cell destruction.

The pattern and timing of queen rearing and cell sealing show great variability, but generally colonies seal an average of 15 to 25 queen cells prior to and immediately after swarming. Colonies most commonly swarm the day of or a day after sealing the first queen cell, usually 8–10 days after queen rearing commences. By swarming at or after queen cell sealing, colonies are ensured of having at least one virgin queen emerge after swarming. But weather and internal colony factors can alter the timing of swarming dramatically, so that colonies may swarm when only very young larvae are present in queen cells or after a mature virgin queen has emerged.

Queen cell destruction prior to swarming delays the issuing of the first swarm for a few days, a delay that may improve the chances of swarm or parental colony survival. For example, queen cells are frequently destroyed during periods of poor weather, delaying swarms from issuing under cold or rainy conditions. For swarming to succeed colonies must time their swarming efforts to coincide with a period of growth in the adult population and the presence of enough brood in the colony to compensate for the loss of adult workers in swarms. It is notable that queen rearing begins precisely when worker brood rearing is at its peak, and there are almost no unoccupied cells in the central brood area (Winston and Taylor, 1980; Winston, Dropkin, and Taylor, 1981) (Fig. 11.2). Thus, colonies time their queen rearing to coincide with population peaks, and queen cell destruction may serve to ensure that colonies do not swarm until there are sufficient adults and brood to populate swarms and the original colony. Other explanations for queen cell destruction may be that poorer quality queen larvae are destroyed, or that sister kinship groups within the colony destroy developing queens of their half-sisters; the latter is discussed in the next chapter.

Other aspects of worker and queen behavior begin to change prior to swarming. The queen is fed more frequently and lays more eggs until the week before swarming, when workers feed her less, her egg laying decreases, and her abdomen diminishes in weight so that she can fly with the swarm (Taranov and Ivanova, 1946; Allen, 1955, 1956, 1960). The workers also shake, push, and bite the queen, generally treating her roughly, which forces her to keep moving. The queen examines the queen cells frequently, putting her head in the cells, sometimes damaging unsealed queen cells in which she may have earlier laid eggs. The workers become relatively quiescent a few days prior to swarming, and clusters can be seen festooned quietly on the bottom of combs. Scouting for new nest sites may begin at this time (Lindauer, 1955).

The peace and quiet of a preswarming colony changes dramatically on the day of swarming, when colonies erupt in the seemingly unorganized and frenetic activity which leads to a swarm. Swarms generally issue in the late morning or early afternoon, although daily weather conditions affect the time of issuance so that swarms can occur at almost any time during daylight hours. Because of the un-

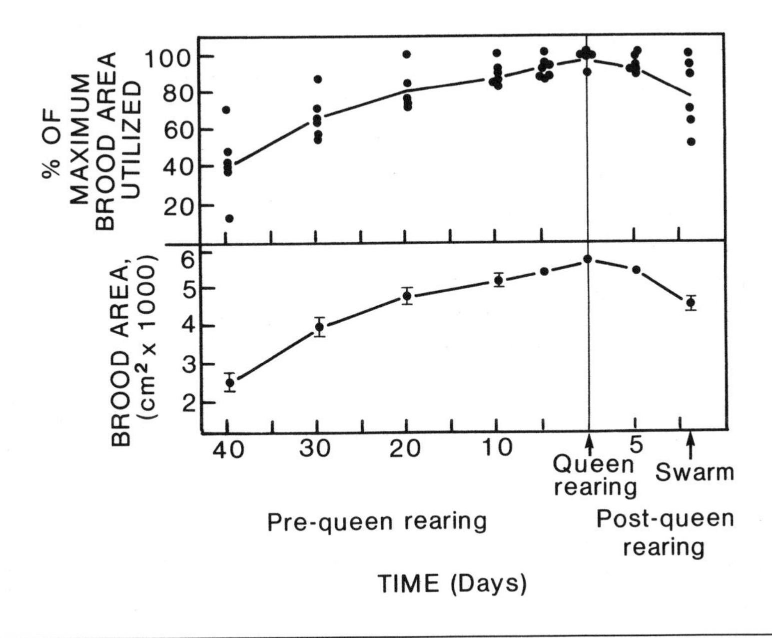

Fig. 11.2

The proportion of the brood area utilized and the total amount of worker brood in colonies preparing to swarm. (Redrawn from Winston and Taylor, 1980. Used by permission of MASSON S.A., Paris.)

predictability of the day and time when swarming will occur, workers engorge with honey for about 10 days prior to swarming, to ensure that they will be carrying sufficient honey reserves when the swarm finally issues; queen rearing is probably the stimulus for engorgement. Both workers which swarm and those which stay behind carry an average of 36 mg of honey in their honey stomachs, as opposed to the 10 mg found on average in stomachs of workers in nonswarming colonies (Combs, 1972). This extra load of honey, which is approximately 40% of the total worker weight, provides food reserves for the swarm while in transit and during the first few days at the new nest.

Workers abruptly change their behavior during the few hours which precede swarming (Allen, 1956; Fletcher, 1975). The rate of dorsoventral abdominal vibrations suddenly drops (see Fig. 9.6), and workers begin running back and forth in waves, buzzing to excite the other workers (Esch, 1967). The queen is chased, bitten, and pulled along with the excited workers. Suddenly, a torrent of workers pours out of the nest entrance and takes to the air, usually driving the queen out also. Occasionally, "false swarms" issue, when the queen is not driven out or reenters the nest. In that case the clustering workers become nervous, take flight, and return to the colony. If the queen joins the forming cluster, the remaining flying workers orient to her and the cluster, and once the cluster has fully formed scouts fly out searching for potential nest sites. This first swarm with the old mated queen is referred to as the prime swarm.

Worker age is the major factor determining which workers will remain in the nest and which will issue with both the prime swarm and subsequent casts called "afterswarms" (Butler, 1940; Meyer, 1956b; Winston and Otis, 1978). Younger workers have a higher probability of issuing with swarms than older workers, and up to 70% of workers less than 10 days old leave with swarms of temperate-evolved bee races (this figure is higher for tropical bees, see Chapter 14). The advantage of a higher proportion of young workers issuing with swarms is that it provides more bees with greater potential longevity for the swarm, a factor of critical importance since new workers will not begin emerging until at least 21 days after the swarm has colonized a new nest site. In contrast, the original colony contains emerging brood which provides an influx of young workers for the old nest.

Another factor that may influence which workers issue with swarms is kinship. In one study preliminary evidence suggested that workers tend to go with queens to which they are related as full sisters rather than half-sisters. That is, workers issue with a queen which had the same father as they did (Getz, Bruckner, and Parisian, 1982). Subsequent research failed to confirm this hypothesis, when workers of two different patrilines segregated randomly during swarming (Getz and Winston, unpublished observations). Also, some drones may issue with swarms, although the number is generally small, about 1% of the total swarm population (mean, 151 drones, range, 0–1001; Avitabile and Kasinkas, 1977).

Once the swarm has found a new nest (see Chapter 5), it lifts off and moves to colonize the site. Buzzing runs by scouts trigger the move, and in less than a minute the swarm forms a circular cloud about 10 m in diameter and 3 m high. The swarm accelerates to about 11 km/hr, and scouts "pilot" the swarm by performing streaking flights through the flying bees. The swarm halts at the new nest, and scouts drop to the entrance and release Nasonov pheromones to attract the swarm to the nest entrance (Ambrose, 1976; Seeley, Morse, and Visscher, 1979).

At this point swarming is not yet completed for the original colony, which still contains unsealed and sealed queen cells, worker brood, some stored honey and pollen, and approximately 40% of the adult workers. After the prime swarm has issued, workers continue to seal queen cells and rear the remaining worker brood, so that the colony population rebounds somewhat as young workers emerge (Fig. 11.1). The worker brood mortality rate is high during this period, about

42%, with most mortality occurring in the egg and young larval stages (Winston, Dropkin, and Taylor, 1981). New virgin queens begin to emerge about a week after the prime swarm issues, although queen emergence is variable depending on the timing of prime swarming relative to cell sealing (Otis, 1980).

Once virgin queens mate, colonies either produce afterswarms or kill all but the one remaining queen which will mate and reign over the nest. In the most typical scenario the first virgin queen to mature issues with an afterswarm 2–4 days after her emergence, and often a second afterswarm issues a few days later with the next queen to emerge (Winston, 1979c, 1980; Otis, 1980; Winston, Dropkin, and Taylor, 1981; Lee and Winston, unpublished observations). Anywhere from 0 to 4 afterswarms can issue from nests, with means between 1 and 2 afterswarms per colony recorded for temperate-evolved bee races. The worker population in swarms generally decreases with swarm number; in one study prime swarms and first and second afterswarms had means of 16,000, 11,500, and 4000 workers, respectively (Winston, 1980) (Fig. 11.1). Third and fourth afterswarms frequently contain only a handful of workers. Swarms can be much larger or smaller than these values, however; as few as 1750 and as many as 50,750 workers have been found in swarms (Avitabile and Kasinkas, 1977). There is a strong correlation between the area of sealed brood when the prime swarm issues and the number of afterswarms. Thus, colonies regulate the extent of their afterswarming according to the number of workers required to remain in the original colony.

Afterswarms are similar in most respects to prime swarms, except that virgin queens participate rather than mated queens. Although only one virgin queen usually issues with an afterswarm, other mature queens from cells sometimes escape during the chaotic period when afterswarms issue, and afterswarms can contain up to three queens (Avitabile and Kasinkas, 1977; Otis, 1980). Workers in afterswarms do not engorge as fully as in prime swarms, and fewer young workers issue with afterswarms (Winston and Otis, 1978; Otis, Winston, and Taylor, 1981).

There is considerable interaction between virgin queens and queens and workers following the first queen emergence. Many of these behaviors are designed to prevent the emergence of more than one queen at a time and also to regulate fights between queens when swarming is completed. The presence of a virgin queen in a colony usually suppresses the emergence of other queens, probably by stimulating dorsoventral abdominal vibrations by workers on queen cells containing mature queens (see Chapter 9). An emerged queen announces her presence by pheromones and piping—a series of pulsed, high-pitched sounds produced by a queen pressing her thorax against the comb and operating her wing-beating mechanism without spreading her wings (Simpson and Cherry, 1969). Mated queens sometimes pipe before swarming (Allen, 1956), but the frequency of piping is greatest between the time the first virgin queen emerges and the end of afterswarming, when the remaining queens fight. The direct effect of piping is to cause workers to freeze in place on the comb until the piping is completed. When a piping queen is present, workers will not chew away the wax and fibers on the capped end of queen cells, thus preventing queen emergence. Piping may also serve to inform the workers and the emerged queen that other queens are present, since mature queens inside cells also pipe. This information is important to the colony, because the original colony will become queenless if it swarms in the absence of any developing queens.

Occasionally, two or more virgin queens are released at once, and these queens may tolerate each other's presence for a few hours or days before fighting. Once afterswarming is completed, emerged queens attempt to kill queens held inside the cells by cutting small holes in the cell sides and stinging the occupants. Workers may then complete cell destruction and destroy the queens inside whether or not they have been stung by the emerged virgin (Caron and Greve, 1979).

When two virgin queens emerge simultaneously, they fight to the death, and the injured combatant in these fights is finished off by the workers. Fighting queens attempt to sting and chew each other with their mandibles, and workers form a ball around the unsuccessful queen in order to kill her. Older workers tend to be involved in the balling behavior more than young ones (Robinson, 1984), and the queen fights and worker balling continues until only one queen remains alive in the nest. At that point the queen begins her mating flights and, once mated, initiates egg laying. The entire swarming period from the initiation of queen rearing until the new queen begins egg laying takes about 4 weeks.

Reproduction by swarming is a risky process, and it is difficult to understand since relatively few swarms survive in cold temperate climates. In a study done in central New York state only 24% of colonies founded from swarms survived until the following season (Seeley, 1978). Only 8% survived in a study in Ontario (Morales, 1986); a similar two-year study in British Columbia found that none of 30 swarms caught and hived during the spring survived the subsequent winter (Lee, 1985). The mortality rate was lower for established colonies which had survived one season, with 78 and 45% of those colonies found alive the next season in the New York and Ontario studies, respectively. Once the New York colonies survived the initial season, mean colony longevity was 5.6 years. Thus, while an individual swarm has a low probability of survival, those which do survive the first season may persist for many years.

Given the low probability of swarm survival, it is not surprising that there are strong correlations between swarm size, timing, growth, and survival characteristics (Lee and Winston, 1985a,b; Lee, 1985; Seeley and Visscher, 1985). Larger and earlier swarms survive longer than smaller, later swarms and also show better initial growth. For example, significant positive relationships exist between swarm size and the rate of comb construction and brood rearing, while negative relationships are found between the date of swarm establishment and these factors. In addition, emergent workers weigh more in colonies established from larger swarms, suggesting that smaller swarms are constrained not only in the number of workers they can initially rear, but also in the quality of those workers.

Under these constraints it might be expected that annual colony growth and reproductive characteristics would favor large, early swarms, and this prediction is supported by some colony traits. First, the allocation of honey and pollen stores to mid-winter brood rearing results in colony population growth which permits early spring swarming. Second, prime swarms are large, containing 60% or more of the adult population at the time of swarming, which greatly increases their survival potential. Third, workers issuing with swarms engorge heavily with honey, and many young workers issue, both factors encouraging comb building and brood rearing to begin almost immediately after the establishment of a new nest.

But extensive afterswarming does not appear to benefit colonies or their swarms, particularly second to fourth afterswarms, since they are small, issue relatively late, and thus have reduced chances of surviving. Furthermore, afterswarms have virgin queens which must successfully mate and return to the new colony, and colonies do not begin brood production until the queen begins egg laying a week or more after nest establishment.

Nevertheless, there is some evidence that afterswarming is not as maladaptive as it might appear. Colonies only produce as many afterswarms as their brood and adult populations can support; the number of afterswarms is positively correlated with both the amount of brood and the number of adults in colonies at the time of swarming (Winston, 1979c, 1980; Lee, 1985). Also, colonies produce more afterswarms during better years and in locations with longer growing seasons. In Kansas, for example, colonies produced an average of 2.0 afterswarms during one spring, whereas further north, in British Columbia, colonies produced averages of 1.0 and 1.5 afterswarms during two consecutive years (Winston, 1980; Lee, 1985), and Ontario colonies produced only 0.15 afterswarms per swarming cycle (Morales, 1986). Total reproductive rates were 1.0 in Ontario, 1.9 and 2.6 during the two years in British Columbia, and 3.6 in Kansas, reflecting the higher incidence of multiple swarming episodes and afterswarm production in warmer temperate regions. Thus, while afterswarms are at more risk than prime swarms, colonies only swarm as often or produce as many swarms as their size, latitude, and seasonal factors permit without unduly diminishing the fitness of the original colony.

A final point concerning afterswarms is that they are more adaptive in tropical habitats than in cold temperate ones. Many small afterswarms do survive in tropical regions (Otis, 1980), and even in Kansas some of the afterswarms which were hived in new nests survived and swarmed the following spring (Otis, unpublished observations). Thus, the Ontario and New York studies cited above reflect conditions close to the northern limit of the honey bee's range and well beyond the Mediterranean climate where the Italian race of bees (*A. m. ligustica*) used in those studies originally evolved.

The Factors Which Induce Colonies to Swarm

The question of what induces swarming is most accurately studied by examining the factors that induce queen rearing, since, once queen rearing begins, there is a well-defined albeit varied sequence of events which leads to swarming. A considerable amount of research has been devoted to formulating and testing various hypotheses concerning queen rearing and swarming, as swarm prevention is probably the major management problem confronting beekeepers worldwide. Two major hypotheses were proposed in the early phases of swarming research to explain the initiation of queen rearing prior to swarming: one, the nurse bee or brood food hypothesis (Gerstung, 1891; Morland, 1930), states that a surplus of young nurse bees develops in preswarming colonies, resulting in an excess amount of brood food for which queen rearing is an outlet. The other, the colony congestion or crowding hypothesis (Huber, 1792; Demuth, 1921), posits that crowding of adult workers and limited space for brood rearing results in the initiation of queen rearing.

Neither hypothesis, however, has proven adequate to explain the initiation of queen rearing. The only experiments supporting the brood food hypothesis are those of Perepelova (1928a), who found that none of four colonies began queen rearing when unsealed brood was added, thereby increasing the demand for brood food. In contrast, most other experiments designed to produce excess brood food, either by removing unsealed brood (Demuth, 1922, 1931; Ribbands, 1953), adding emerging sealed brood, or both (Demuth, 1931; Simpson, 1957) have failed to induce swarming. Moreover, the proportion of nurse bees to brood increases as summer advances, peaking in autumn (Bodenheimer, 1937), and the ratio of young bees to unsealed brood increases most dramatically after queen rearing is initiated (Winston and Taylor, 1980), rather than before, as would be predicted by the brood food hypothesis.

In testing the crowding or congestion hypothesis, experiments designed to restrict hive space have resulted in queen rearing and swarming in many but not all colonies (Perepelova, 1947; Simpson, 1957, 1973; Simpson and Riedel, 1963; Simpson and Moxley, 1971). However, approximately half of the uncongested colonies from studies by Allen (1956), Gary and Morse (1962), and Simpson (1973) also initiated queen rearing, and restricting the number of cells available for oviposition has not always led to swarming (Simpson and Riedel, 1963; Simpson and Greenwood, 1975). Thus, although limited nest space, restricted numbers of cells, and congestion may play a role in stimulating swarming preparations, none of these factors alone consistently induces swarming.

One difficulty with these hypotheses is that they include only one factor, whereas queen rearing and swarming are extraordinarily complex functions involving well-timed and coordinated activities by thousands of individuals. It is more likely that there are multifactorial cues for the initiation of queen rearing, based on certain within-colony demographic factors which not only stimulate queen production but also contribute to the success of swarming (Simpson, 1958; Winston and Taylor, 1980; Lensky and Slabezki, 1981). That is, queen rearing coincides with a short "window" in time during which colony conditions are most favorable for swarm production, and most of these colony characteristics must be at or near their threshold levels for queen rearing to begin. The relationship between colony conditions and queen rearing can be summarized as follows (Fig. 11.3): Queen rearing is initiated due to intrinsic (demographic) and extrinsic (resource abundance) factors inducing workers to begin rearing new queens at a time when conditions are favorable for swarm production. The primary stimuli, none of which would initiate queen rearing independently of others, include (1) colony size, (2) brood nest congestion, (3) worker age distribution, and (4) reduced transmission of queen substances. Resource abundance influences the first three factors and also may be a primary stimulus for queen rearing.

The evidence supporting this hypothesis is presented below and is based in part on the remarkably low variability found in certain colony characteristics when queen rearing begins. Some supporting

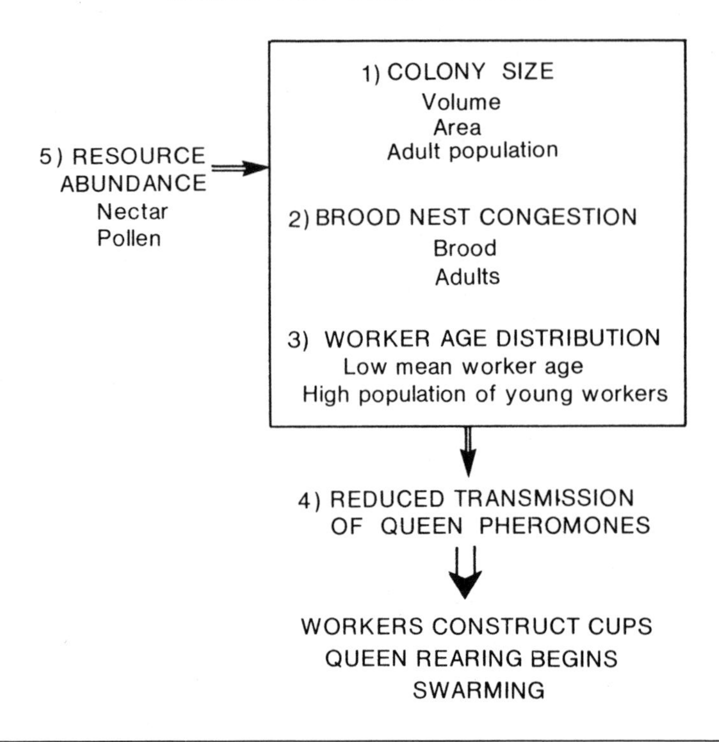

Fig. 11.3

The stimuli which induce workers to begin queen rearing.

data for tropical-evolved Africanized bees in South America and temperate-evolved European bees in North America are presented below. (For a more complete data presentation and discussion, see Winston and Taylor, 1980, and Winston, Taylor, and Dropkin, 1981.)

COLONY SIZE

Three aspects of colony size are important for the initiation of queen rearing: comb area, colony volume, and worker population. Both Africanized and European bees swarm when colonies have reached a certain size, about 20–25 L and 8000 cm^2 of comb for Africanized bees, and 40 L and 23,000 cm^2 of comb for European bees. In both races queen rearing begins when there are about 12,000 workers in the nest, and colonies swarm when the worker population has reached approximately 20,000 workers (Fig. 11.1). Thus, queen rearing begins during a period of accelerated growth in the worker population, but before colonies reach their maximum populations.

The most important aspect of colony size is not the physical size of the nest, however, but the size of the active colony, and this factor may be expressed in slightly different ways for tropical and temperate bees. Tropical-evolved bee races initiate queen rearing at the size given above, regardless of the potential volume of their nest (Fig. 11.4) and while utilizing 80–86% of the comb (Fig. 11.5). Thus, their active colony size at swarming is close to that of the total nest size. In contrast, temperate-evolved European bees may construct enough comb to fill their cavities before swarming, although only 54–76% of that comb is used when queen rearing begins (Fig. 11.5).

BROOD NEST CONGESTION

Two aspects of brood nest congestion may be important stimuli for queen rearing: congestion of brood and crowding of adult workers. The effective brood nest can be described as that with the maximum number of cells occupied by brood during a swarming cycle, and queen rearing is initiated coincidental with congestion in that area. At that time 90–95% of all cells in the brood nest contain some stage of brood, and there are few empty cells available for oviposition.

When queen rearing begins, the number of workers is rapidly increasing (Fig. 11.1), and clusters of quiescent bees can be found on comb containing brood. Many of the brood combs are covered by layers of workers two to three deep, although peripheral combs do

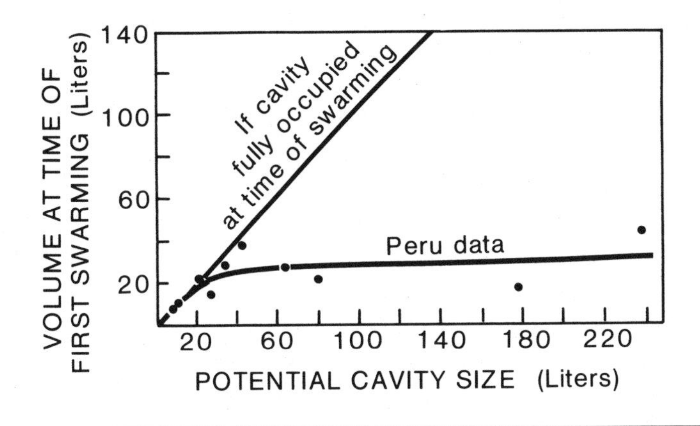

Fig. 11.4

Cavity volume occupied by ten feral Africanized bee colonies at the time of first swarming near Pucallpa, Peru, versus the total volume of the cavities. The straight line indicates potential cavity size at first swarming if the cavities had been completely full. (Redrawn from Winston and Taylor, 1980. Used by permission of MASSON S.A., Paris.)

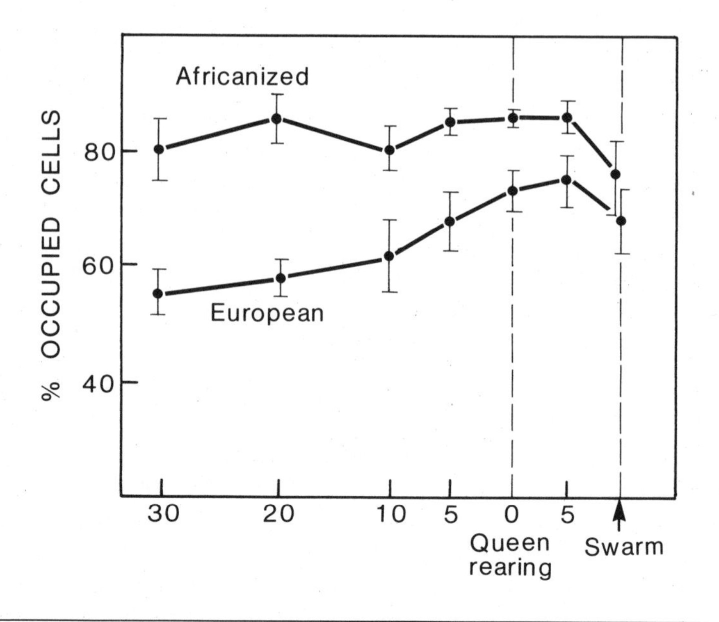

Fig. 11.5

Mean percentages and standard errors of total comb area occupied by brood and stored honey and pollen for Africanized bees in South America and European bees in North America. (Redrawn from Winston, Dropkin, and Taylor, 1981.)

not have as high a worker density. Thus, crowding of both brood and adult workers in the brood nest area is more important than overall crowding in the nest cavity.

WORKER AGE DISTRIBUTION

The age distribution of workers is also important for the initiation of queen rearing, with colonies characterized by low mean worker age and high proportions of young workers when queen rearing begins. For example, close to half of the workers in colonies are less than 8 days old when queen rearing starts. Colonies begin to show this age distribution approximately 15 days prior to the initiation of queen rearing, although with a lower adult population. Thus, worker age distribution may only be important when sufficient workers are present for queen rearing to begin.

REDUCED TRANSMISSION OF QUEEN SUBSTANCES

Another factor in swarm preparations is a reduction in the transmission of queen pheromones which inhibit the rearing of new queens by workers (reviewed in Chapter 8). There are no differences in the production of one queen pheromone, 9ODA, between queens in colonies preparing to swarm and those not preparing to swarm (Seeley and Fell, 1981), which suggests that it is transmission of queen pheromones which is reduced rather than the queen's pheromone output prior to queen rearing. Reduced dispersal of queen pheromones by messenger bees is one consequence of colony congestion and partly explains the removal of the pheromone-mediated inhibition of queen rearing. As colonies become crowded and workers are more quiescent, the queen's movements in the nest may also be diminished, further reducing pheromone dispersal and limiting the amount of footprint substances she can distribute over combs, which may in turn inhibit queen cup construction (Lensky and Slabezki, 1981).

RESOURCE ABUNDANCE

Factors extrinsic to the colony, such as availability of nectar and pollen, also exert some influence on the initiation of queen rearing. The indirect effects of resource abundance are obvious, since colonies cannot grow to the proper size and level of congestion to swarm in the absence of good nectar and pollen sources in the field. The direct influence of this factor is more difficult to determine, although it is possible that an innate seasonal cycle may predispose workers to swarm at peak resource availability (Morland, 1930; Simpson, 1958). This is somewhat unlikely, however, since bees transplanted from one region to another do not swarm at the same time of year as in the location where they evolved; rather, the swarming occurs during periods of more intense flowering. Furthermore, seasonal effects do not explain the small peak in swarming in late summer (Fig. 11.1). Thus, the effect of resources is probably an indirect rather than a direct seasonal stimulus for queen rearing to begin.

The colony demography hypothesis requires that workers be able to perceive colony size, congestion, age structure, pheromone levels, and possibly resource abundance as well. Although a mechanism for perception of such colony conditions has not been demonstrated in the context of swarming, there are a number of ways that features of colony demography might provide relevant information leading to queen rearing. For example, Seeley (1977) has shown that scouts searching for nesting sites use a process akin to integral calculus to measure cavity sizes; awareness of colony size and patterns of comb utilization may be possible by a similar process. Perception of colony age structure as well as the amount of brood may also be possible, by such mechanisms as extensive patrolling (Lindauer, 1952), transfer of nectar and pollen between workers (Free, 1965), and amount of brood feeding necessary. Workers may be aware of the number of sealed queen cells by the likelihood of encountering such cells while patrolling. Brood nest congestion could be perceived by the number of workers and empty cells encountered while moving through the brood area, or by density-dependent changes of hormone levels. Mechanisms for perception of resource abundance are better known, and include time required to locate resources, dancing patterns in the colony, and transfer of nectar between workers. Colony conditions may also influence the age of workers issuing with swarms; Taranov (1947) found that the percentage of young workers remaining in the nest at swarming was correlated with the amount of brood present.

The indirect effects of crowding and congestion might also provide cues to workers which stimulate queen rearing. Reduced dispersal of queen pheromones is certainly one such effect, but environmental conditions within the nest are also altered prior to queen rearing and swarming. Increased temperature and reduced ventilation are associated with congestion prior to swarm preparations (Lensky and Siefert, 1980), and worker perception of these changes in nest environment could stimulate new queen production.

The relationship between colony characteristics and the initiation of queen rearing is thus one in which thresholds of many demographic characteristics must be reached before swarm preparations begin. This multifactorial, threshold-controlled concept may explain the failure of many studies to completely induce or inhibit swarming by single-factor manipulations. For example, not all colonies swarm when nests are artificially crowded, and some colonies do swarm when nests are given substantial extra comb and/or cavity volume (Murray and Jeffree, 1955; Simpson, 1957, 1958, 1973; Gary and Morse, 1962; Simpson and Riedel, 1963; Allen, 1965; Simpson and Moxley, 1971; Simpson and Greenwood, 1975; Caron, 1981). Another example is that queen rearing can start and stop frequently before swarming actually occurs (Allen, 1965a; Gary and Morse, 1962), which may result from some swarm-inducing factors vacillating at or slightly below their threshold levels. Finally, swarming in late summer (Fig. 11.1) may result from some colonies reaching their demographic thresholds at a time of year when swarming is clearly maladaptive but proceeding to swarm since colony conditions are at their appropriate levels.

A compelling aspect of the role of within-colony demographic factors in stimulating swarm preparations is that the cues involved allow workers to time swarming so as to coincide with colony conditions that maximize the chances for successful swarm production and maintenance of the original colony. Factors such as a worker age distribution skewed toward young workers, sufficient brood and adult populations to support swarming, and early swarm production are critical for swarm and colony survival, and these are some of the same factors which workers may use directly or indirectly as cues for queen rearing. Fruitful areas of future honey bee research will certainly include a closer examination of the relationship between preswarming conditions in colonies and the success of their offspring, as well as manipulations designed to test a multifactorial origin for the initiation of queen rearing. With this approach, a full understanding of swarming should be within our grasp.

Supersedure

There is another type of queen reproduction in honey bees, supersedure, in which the old queen is eliminated. Generally, no swarms are produced when colonies supersede their queen, but when a swarm does issue it contains a new virgin queen rather than the old, mated queen (Alfonsus, 1932; Cale, Banker, and Powers, 1946).

The immediate cause of supersedure is probably diminished pheromone production by an older queen (Butler, 1957a), but the cause of a shortage in queen-produced substances is not always clear. Colonies supersede queens which are injured (Wedmore, 1942; Cook, 1968), diseased (Farrar, 1947; Furgala, 1962), or laying unfertilized eggs or an insufficient number of fertilized eggs, but a relationship between these queen problems and pheromone production has not been established. Older queens are superseded more frequently than younger ones, possibly because of age differences in queen pheromone production.

Colonies which are superseding their queens tend to do so in the late spring or early summer, although supersedure can occur anytime from early spring through the fall. In one study about 20% of the colonies superseded their queens in a given year (Allen, 1965a). When superseding, usually fewer than six queen cells are reared, and destruction of some cells during supersedure is similar to that which occurs during swarming. The old queen may continue to lay eggs while new queens are developing, and often she is not eliminated until the successful virgin queen has mated and begun her own egg laying. This tolerance of the old queen presumably is due to her diminished pheromone production but also has considerable adaptive value, because it ensures the presence of a laying queen even if the virgin queen fails to return from her mating flight.

Colony size may influence whether a colony swarms or supersedes its queen; large colonies are more likely to replace their queens by supersedure than by swarming. Morales (1986) found that 50% of colonies kept in 84-L hives, about twice the size of an average feral nest, superseded their queens, and none of those colonies swarmed. In contrast, only 5% of colonies in 21- or 42-L hives superseded their queens, and 80% of those colonies swarmed—100% of those in the

21-L hives, and 60% of those in the 42-L hives. The lower swarming and higher supersedure rates of large colonies may reflect a strategy of delaying reproduction until the following year, thereby increasing the likelihood of overwintering success by entering the winter with a young queen.

12

Drones, Queens, and Mating

Swarming and supersedure are only part of the reproductive biology of honey bees; drone bees are necessary for mating, which completes the reproductive cycle. Although drones exist only to mate and perform no other useful functions in the nest, most die before mating, either because they get old or are thrown out of the nest by the workers. The few drones which succeed in mating queens can do so only once, since they die immediately after mating, when their abdomens and genital apparatus rupture. Nevertheless, drones spend most of their adult lives flying out to congregation areas with hundreds or thousands of other drones, competing for the favors of the few queens which make their way to the mating sites.

The structure of this odd mating system can best be understood in the context of the social structure of honey bee colonies. The timing and extent of drone production by colonies is controlled by many of the same factors involved in swarming. Drones are produced and maintained only when colonies can support them and when queens are potentially available for mating. The mating system is designed so that queens can mate with many drones, most often with drones from other nests.

The Premating Biology of Drones and Queens

Honey bee colonies are protandrous, so that the peak of drone rearing precedes the emergence of virgin queens in the spring (Allen, 1963; Page, 1981; Lee, 1985) (Fig. 12.1). Because drone rearing generally peaks about 4 weeks before swarming, drones can emerge and mature before virgin queens are available for mating. Colonies breed fewer drones in the summer, when few virgin queens are produced, but there is a slight rise in drone rearing in late summer coincidental with the August peak in swarming (see Figs. 11.1 and 12.1). By September, drone rearing has ceased until the following spring.

The number of drones reared by colonies is determined by the amount of drone comb constructed, the colony size, and whether the

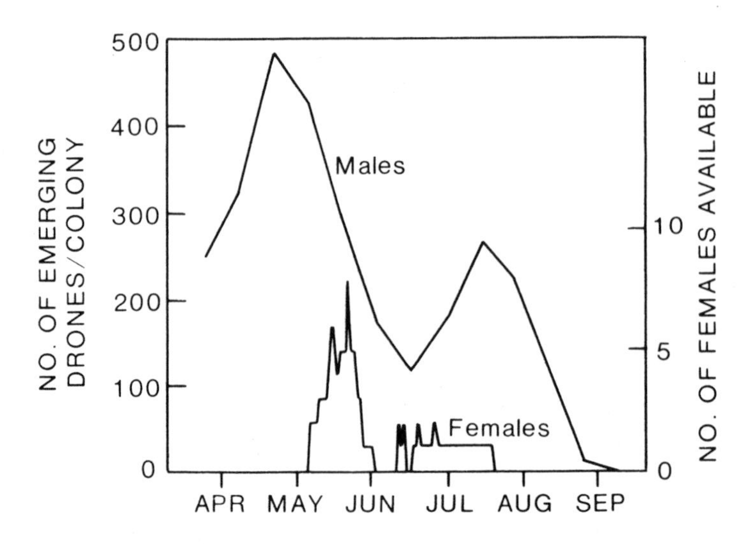

Fig. 12.1

Temporal emergence pattern of drones per colony compared to the number of unmated reproductive females in the population during one year of study in British Columbia. (Data from Lee, 1985.)

colony is an overwintered colony, a prime swarm, or an afterswarm. New swarms begin constructing comb with worker-sized cells almost immediately; drone comb construction begins an average of 22 days after colony founding and is 90% completed for the first season within 64 days of colony initiation. The amount and proportion of drone comb constructed is strongly influenced by swarm size and date of swarming, with large, early swarms producing considerably more drone comb than small, later ones. New swarms construct between 0 and 35% drone comb, with a mean of 8%, depending on these factors (Free, 1967a; Taber and Owens, 1970; Lee and Winston, 1985b). The diminished or complete absence of drone comb construction in colonies founded by smaller swarms can be explained by a colony's resource allocation priorities. Small colonies collect less nectar and construct less comb than large colonies, and since drones do no work and consume colony resources, they are a liability to colony growth. Thus, small colonies would not be expected to produce drone comb until they were larger and had sufficient resources for drone comb construction. Similarly, the 22-day delay in the initiation of drone comb by swarms is due to the priority of worker comb for early colony growth.

Additional drone comb may be constructed by colonies which survive the winter, and most of this comb building takes place in the spring. When colonies have completed all of their drone comb construction, an average of 13–17% of the total comb area is devoted to drone comb (Seeley and Morse, 1976). The proportion of drone comb built by a colony in the spring following nest establishment depends in part on the amount of drone comb already present. Thus, there is a feedback mechanism by which workers can control the amount of drone comb they construct, thereby partially controlling the number of drones reared. Small overwintered colonies also produce a smaller amount and proportion of drone comb than do larger colonies (Free, 1967b; Free and Williams, 1975).

The number of drones reared also depends on colony population and type. In the spring average-sized feral colonies rear between 660 and 3960 drones prior to swarming (mean, 2400), and an additional 0 to 1600 drones before the fall (mean, 550). Larger-sized colonies typical of those used for beekeeping, and provided with unlimited amounts of drone comb, can rear many more drones than feral colonies, up to 45,000 in a season (Allen, 1965b; Page, 1981; Page and Metcalf, 1984). Newly established swarms rear between 0 and 8200 drones their first season, with prime swarms rearing a mean of 3690 drones and afterswarms a mean of 830 drones (Lee, 1985).

More populous colonies or swarms rear more drones, and the majority of drone rearing occurs early in the season, as do other effects of swarm size and timing on drone comb construction (Free and Williams, 1975; Lee, 1985). Thus, it is not surprising that afterswarms rear fewer drones than prime swarms, since they issue later and are smaller. Similar factors may explain the more extensive drone production of prime swarms compared to established colonies following swarming, since prime swarms have more workers than colonies that have swarmed and begin brood rearing immediately, in contrast to swarming colonies, which generally afterswarm and must then wait for a new queen to mate before brood rearing can begin. Because few swarms survive the winter, the extensive drone production of prime swarms may reflect an attempt to reproduce through their drones mating with supersedure or swarm queens prior to the onset of winter.

Although the influence of drone comb area and colony size and type on the extent of drone rearing seems well established, the factors which induce workers to initiate drone comb construction and drone brood rearing are not as clear. The initiation of drone rearing prior to swarming suggests that factors similar to those which induce swarming are involved, though at different levels. Colonies may be able to track the timing of swarming in the general population according to their own demographic condition and begin drone rearing during the spring period of colony expansion. Pheromones may also be involved, as they are in the initiation of queen rearing (Free and

Williams, 1975). But there is no good evidence favoring either of these hypotheses, and our knowledge of the factors inducing drone rearing remains speculative.

Once drones begin to emerge, they commence activities and physiological changes which prepare them for mating. Drones are initially fed by workers for the first few days of their lives, but gradually they begin to feed themselves from honey cells (see Chapter 4). The development of drone reproductive organs continues for about 12 days after emergence, when drone semen has matured sufficiently for mating (Ruttner, 1966). During this early adult period, drones tend to be found in the center of the nest, on comb containing brood, where temperatures are warmer and nurse bees more available to feed the young drones (Free, 1960a; Ohtani and Fukuda, 1977). As drones age, they are more frequently found on peripheral combs where honey is stored or near the nest entrance. While in the nest, drone activities are mostly limited to resting, begging food from workers or feeding themselves from honey cells, and self-grooming (Ohtani, 1974). Drones sometimes form loose aggregations at comb edges and occasionally touch antennae, but interactions between drones and between drones and the queen are uncommon.

Drones generally are driven out of colonies in the fall or when resources are scarce, both times when colonies cannot afford the energy required for their maintenance (Morse, Strang, and Nowakowski, 1967; Free and Williams, 1975). At these times workers pursue drones and force them out the entrance, although the removal of drones from the nest may take many weeks or even months. Colonies do not always throw their drones out, however; populous colonies with substantial honey storage or those without queens may allow drones to remain through a summer dearth period or the winter months (Fukuda and Ohtani, 1977; Ohtani and Fukuda, 1977).

During the spring and summer months drones begin to take orientation flights when about 8 days old, while still sexually immature (Ruttner, 1966). These orientation flights as well as the actual mating flights generally take place in the afternoon, most often between 2:00 and 4:00, although the timing of drone flights can be shifted according to daily weather patterns (Howell and Usinger, 1933; Oertel, 1956; Taber, 1964; Ruttner, 1966; Lensky et al., 1985; Taylor, Kingsolver, and Otis, 1986). The orientation flights are no longer than a few minutes, but the mating flights, which begin when drones are sexually mature, average about 25–32 min, and up to 60 min, although flights on cold, cloudy days are much shorter (Howell and Usinger, 1933; Oertel, 1956; Witherell, 1971). During sunny weather drones take several flights a day, averaging three to five flights an afternoon and returning to the nest for about 15 min between flights to consume additional honey for energy (Howell and Usinger, 1933; Witherell, 1971). Before flying, drones hesitate at the entrance and groom, paying special attention to the antennae, which are carefully cleaned with the antenna cleaner, and the eyes. They then leave the colony to search for congregation areas where mating may occur.

Honey bee queens only mate during the early period of their lives, prior to the commencement of egg laying (reviewed by Ruttner, 1956b; Hammann, 1957; Woyke, 1964, and references cited therein). Before virgin queens become sexually mature, at about the fifth or sixth day after emergence, the workers do not pay much attention to them. Once a queen is ready to mate, however, workers begin to form a court around her, being particularly attentive in the afternoon when mating flights occur. She may be treated roughly by the workers, who vibrate her up to 1200 times/hr, pull at her wings, and sometimes cling to her legs. Abdominal contractions are apparent at this time, possibly associated with opening the abdominal orifice for mating. The queen pipes frequently, perhaps to protect herself from the worker's rough handling, since workers near the queen freeze while she is piping. Just prior to a queen flight, a group of fanning bees or "callers" position themselves at the nest entrance, releasing Nasonov pheromones to assist her in orienting back to the nest. When the queen appears at the entrance, she may attempt to reenter the nest but is usually forced back out by the workers until she flies. While she is gone, the nest entrance remains densely populated with workers, which meet her upon her return.

The orientation and mating flights of the queen occur at about the same time of day as the drone flights, most often in mid-afternoon on sunny days and during periods of improved weather on afternoons with poor weather conditions. Most queens take one or two short orientation flights, and between one and five mating flights, all over a 2-to-4–day period under good weather conditions. Queens may make two or three flights in a day, within a space of 90 min. The number of flights probably depends on both the weather and how successful each flight is; there is some evidence that queens continue mating flights until their spermathecae are full. The desirable weather conditions for mating include temperature above 20°C, little cloud cover, and winds less than 20–28 km/hr. Although some mating can occur under conditions below these thresholds, queens mated during poor weather are frequently superseded, probably because of inadequate transfer of semen during mating. Queens can postpone mating flights for up to 4 weeks after their emergence if weather conditions do not permit flight, but beyond that time they begin to degenerate and lay drone eggs.

Congregation Areas, the Mating Sites

Mating between drone and queen honey bees occurs at congregation areas, discrete aerial sites where drones fly, anticipating the arrival of virgin queens. Mating almost invariably takes place at or near one of these drone congregation areas, while a drone and queen are airborne, and has been difficult to study owing to the speed and height at which it occurs. Although the existence of congregation areas had been known for many years, it was not until the advent of tethering techniques for queens and high-speed cinematography that the details of the mating process could be examined, and there are still many unanswered questions about the location of these mating sites, distances drones and queens fly to mate, orientation mechanisms to the congregation areas, and selection of drones by queens for mating.

The location of congregation areas relative to apiaries or nests can be mapped by flying virgin queens from a balloon or kite and noting the sites where drones are attracted. One notable result of this approach is that there are many congregation areas within flying distance of apiaries (Zmarlicki and Morse, 1963; Ruttner and Ruttner, 1966; Bottcher, 1975) (Fig. 12.2). Congregation areas are not found any closer than 90–120 m from an apiary site, and even these do not attract as many drones as those further away. Thus, drones and queens have numerous options for mating sites, and they tend to mate away from their own nests, thereby increasing the likelihood of outbreeding.

Mapping congregation areas is useful in indicating potential mating sites, but it does not explain why they are found at particular locations or why congregation areas persist at the same site year after year. In one case the same sites in the Austrian Alps were used with virtually no changes for 12 years (Ruttner, 1962; Ruttner and Ruttner, 1972). This is remarkable considering that most drones die during the winter, and thus young, inexperienced drones must find the same areas each spring without prior experience. Moreover, these sites have distinct aerial boundaries, so that a queen just a few meters outside the congregation area will be almost totally ignored by drones (Ruttner and Ruttner, 1965). This suggests that there are some physical characteristics which define congregation areas, but these are not obvious from the sites chosen. Congregation areas tend to be located away from high trees or hills, at sites with at least a hectare of open ground and somewhat protected from the wind, possibly at depressions in the horizon (Zmarlicki and Morse, 1963; Ruttner and Ruttner, 1966). Many congregation areas, however, are found at locations which do not show these characteristics, such as sites over water and forests.

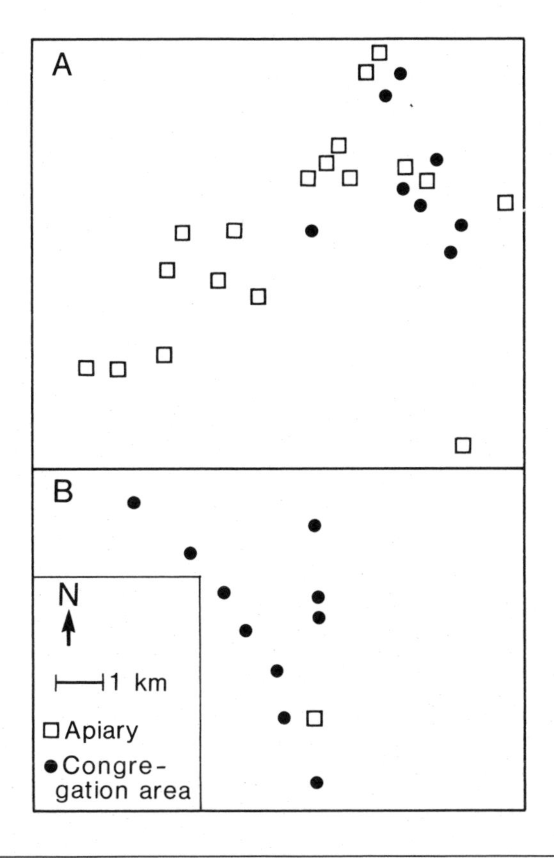

Fig. 12.2

The location of congregation areas relative to apiaries in two different areas: (*a*) derived from data in Ruttner and Ruttner (1966) for a region in Austria; (*b*) derived from Zmarlicki and Morse (1963) for New York State.

It is possible that nonvisual characteristics of congregation areas—for example, localized magnetic or electrical anomalies—may provide orientation cues for drones. A recent study by Loper (1985) found that drones have groups of cells in their abdomens which contain iron-rich granules of magnetite, similar to those found in worker bees which function in orienting to the earth's magnetic field (see Chapter 9). Whereas drones 0–3 days of age have no such magnetite accumulation, drones older than 6 days contain magnetite in these cells, and the young drones may be incapable of directional flight until sufficient iron has accumulated. Based on these findings, Loper has suggested that magnetic sensitivity may be used by drones to locate congregation areas.

One result of the existence of various congregation areas at different distances and directions from any given nest is that there is considerable mixing of drone and queen populations from nests found over a wide area. This population-mixing effect is increased by the tendency of both drones and queens to fly away from their nests to mate, sometimes for considerable distances. Drones from a single nest fly to many different congregation areas, and any one area can be made up of drones from most of the nests within a radius of 5 km or more. An individual drone can visit more than one area during his lifetime and sometimes will attend two or three different congregations in an afternoon (Ruttner and Ruttner, 1966).

There appears to be both a minimum and maximum distance between the source colonies for drones and queens within which mating commonly occurs (Peer, 1957; Ruttner and Ruttner, 1966, 1972; Bottcher, 1975; Ruttner, 1966, 1976; Taylor, Kingsolver, and Otis, 1986). Queens and drones rarely come from colonies less than 2 km apart, and the nests of origin are most commonly separated by 5–7 km. However, matings commonly occur between drones and queens from nests as far apart as 12 km and up to 17 km. Most studies suggest that virgin queens usually fly 2–3 km on mating flights, but there is some disagreement about the average distances flown by drones, which are greater than 2 km in some studies (Ruttner and Ruttner, 1966, 1972; Ruttner, 1976) but less than 2 km in others (Taylor, Kingsolver, and Otis, 1986). Also, while this last study found a steady decline with distance in the numbers of drones found at various congregation areas, the earlier group found no such relationship, at least within a 7-km radius. The differences between these studies may reflect topographical differences between the study sites, or racial and age differences in the drones marked for study, since there is a positive relationship between drone age and the distance drones travel to mate (Taylor, Kingsolver, and Otis, 1986).

The congregation areas themselves vary not only in the characteristics of the surrounding landscape but also in their size and the number of drones in attendance. A typical congregation area contains drones flying in a space 30–200 m in diameter and 10–40 m above the ground (Müller, 1950; Ruttner, 1966; Ruttner and Ruttner, 1966). The number of drones can vary tremendously depending on time of day, weather conditions, and how attractive the area is, and anywhere from a few hundred to many thousands of drones can be found in flight within the boundaries of a congregation area at any one time. One congregation area was estimated to have at least 25,000 drones from more than 200 colonies in attendance at one time (Gary, personal communication, cited in Page and Metcalf, 1982). Drones fly back and forth through the area waiting for queens and produce an audible hum which sounds much like a swarm's. These somewhat lazy flights change to rapid pursuit when a queen arrives, however, and the pursuing drones often assume a comet-shaped pattern as they follow a queen and attempt to copulate.

The factors attracting drones and queens to congregation areas are not known but probably involve physical characteristics of the areas as well as pheromones. The first drones to arrive are thought to be attracted by the physical structure of the surrounding landscape near the congregation areas, although such cues have not been well defined, and artificial congregation areas have been induced with large amounts of queen substance (Strang, 1970). Drones also may release an attractive pheromone (Lensky et al., 1985), so that drones arriving later may orient to pheromones as well as landscape cues. Orientation of queens to the congregation areas may involve similar factors, perhaps with the drone pheromones being more important.

Once a queen arrives at a congregation site and begins to fly through it, drones quickly orient to her using both visual and chemical cues (Gary, 1962, 1963; Gary and Marston, 1971). The drones form temporary swarmlike formations called "drone comets," which form and disintegrate as the drones follow the queen. The drones approach a queen from below, probably because of the dorsal location of their large compound eyes. Approaches from the windward direction of the queen are more successful, presumably because of an odor trail produced by the release of queen pheromones. The initial phases of orientation probably involve pheromones more than visual cues, and visual components may be more important for the final orientation to the queen preceding mounting and copulation. Drones can be attracted by pheromones alone or by strips of black felt which mimic the shape of queens, and models impregnated with pheromones can elicit mating behavior almost identical to that elicited by a real queen (Taylor, 1984a,b).

Mounting and copulation are rapid and spectacular, with the drones literally exploding their semen into the genital orifice of the queen (Laidlaw, 1944; Woyke and Ruttner, 1958; Gary and Marston, 1971; Koeniger, Koeniger, and Fabritius, 1979; Koeniger, 1984). Once contact has been made between the drone and queen, actual mating generally lasts less than 5 sec, and often no more than 1 or 2 sec. Because the queen must open her sting chamber to be receptive to the male, not all drones which mount queens necessarily succeed in mating. As the drone approaches the queen from below, his hind legs hang downward, and in their initial contact the thorax is above the queen's abdomen and the first and second pair of legs straddle the queen (Fig. 12.3). Within a split second the drone grasps the queen with all six legs and everts the endophallus into the queen's open sting chamber (Fig. 12.4). At this point the drone becomes paralyzed and flips backward, and ejaculation results from the pressure of the drone's hemolymph as the abdomen contracts. The explosive and sometimes audible ejaculation ruptures the everted endophallus and propels the semen through the queen's sting chamber and into her

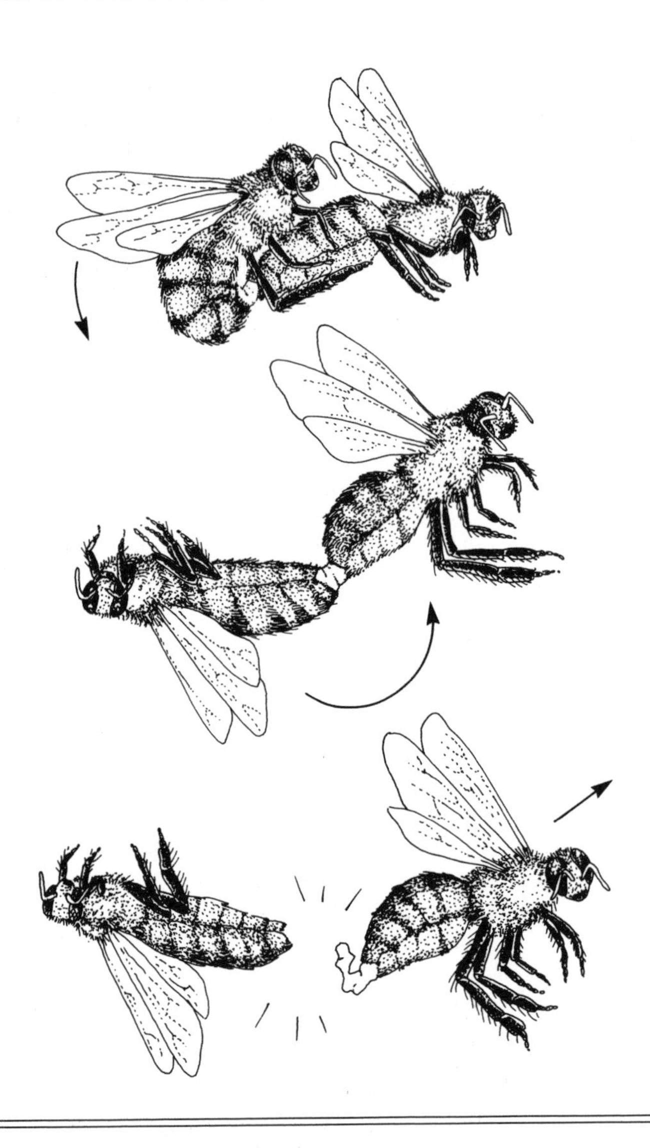

Fig. 12.3

The sequence of copulation events between a drone and a virgin queen in a congregation area.

oviduct. The ejaculation separates the drone from the queen, and he dies within minutes or hours of mating.

In addition to the ejaculation semen, the bulb of the endophallus and/or coagulating mucus are left behind in the queen's vagina, and this plug is called a "mating sign" (Woyke and Ruttner, 1958, and references cited therein). It may function to prevent semen from flowing out of the vagina following copulation but does not prevent subsequent copulations as drones copulating with a queen whose vagina holds a mating sign can push it aside (Bishop, 1920b; Gary, 1963; Koeniger, 1984). A queen returning from a successful mating flight generally is carrying the mating sign of the last drone to mate her, and the workers which greet her lick the sign with their tongues and eventually remove it with their mandibles (Hammann, 1957).

A queen generally mates with more than 1 drone on a single mating flight and an average of 7–17 drones during the few days or weeks she participates in mating (Taber, 1954, 1958; Woyke, 1960; Adams et al., 1977). It is possible for the queen to receive all of her drone semen on one flight; in one instance, a queen seems to have mated with 17 drones on a single flight (Woyke, 1960). Immediately after mating, the oviducts of a queen contain an average of 87 million sperm, and up to 200 million, yet when full the spermatheca contains only 5.3 to 5.7 million sperm, which represents contributions from the ejaculates

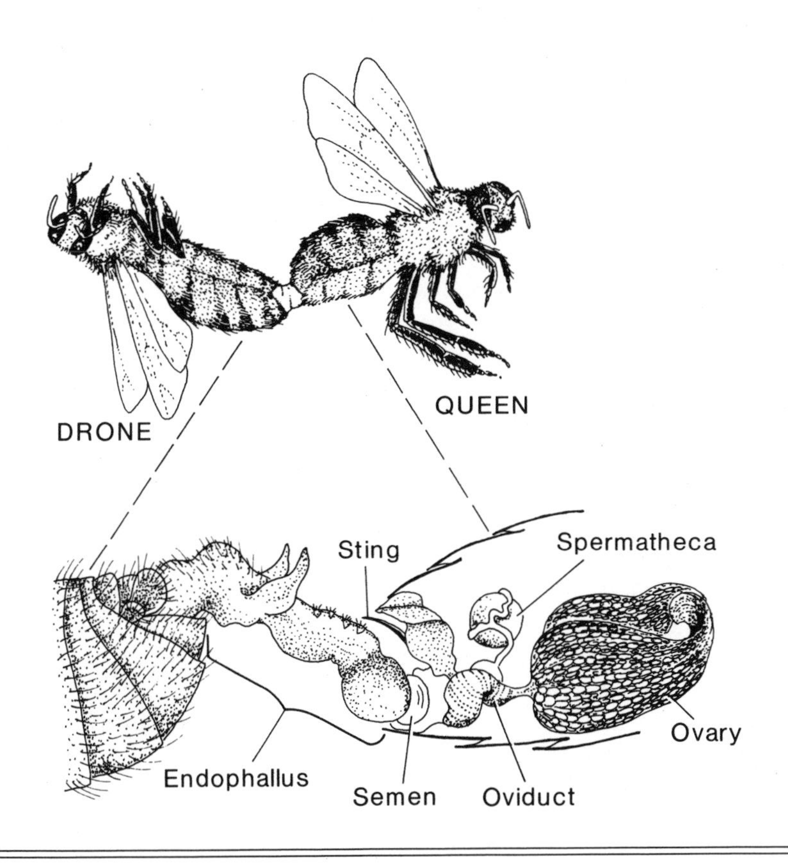

Fig. 12.4

The position of the drone's endophallus in the queen's vagina during copulation. The sperm is being ejaculated into the queen's oviducts as the drone flips backward and begins to separate from the queen.

of all the drones that have mated with the queen (Mackensen and Roberts, 1948; Woyke, 1960; Kerr et al., 1962; Laidlaw and Page, 1984). A queen lays on average fewer than 1500 fertilized eggs daily during the summer, and between 175,000 and 200,000 eggs annually, so that the sperm from these copulations are adequate to fertilize worker eggs for a queen's average life span of less than 4 years (Merrill, 1924; Nolan, 1925; Bodenheimer and Ben-Nerya, 1937; Sakagami, 1958).

The loss of the excess sperm can be explained by the process of sperm migration from the oviducts into the spermatheca, which takes approximately 40 hr (reviewed by Page, 1986). Their migration is driven by abdominal contractions as well as the functioning of muscles attached to the oviducts. After insemination, most of the sperm is packed into the lateral oviducts, having passed beyond the spermatheca, and the sperm moves backward past the valve fold toward the vagina (see Fig. 3.22). Some of this sperm moves actively into the spermatheca through the spermathecal duct, but most of the excess sperm is pumped past the duct and valve fold into the vagina and eventually passes out the sting chamber.

The other puzzling aspect of insemination is the mixing of drone sperm, since sperm from each drone remain in the spermatheca and are released when the queen begins laying eggs. Evidently, sperm does not mix to any great extent in the oviducts, and the rate of sperm entry into the spermatheca is fairly constant. Thus, the different drone ejaculates have fairly equal representation in the spermatheca, although as the spermatheca fills the last sperm to enter may be more poorly represented because of diminished space. Once in the spermatheca, the ejaculates of the various drones mix to some extent, so that the genotypes of all the drones which mate the queen are represented in her offspring workers during her lifetime, and most drones are represented at any one time. Sperm mixing is not complete, however, and there are fluctuations in sperm use which reflect the amount of sperm deposited by each inseminating male. But these fluctuations do not result in overrepresentation or numerical dominance of the sperm or offspring from any one male, so that no sperm dominance or precedence occurs (Page and Metcalf, 1982; Woyke, 1983; Laidlaw and Page, 1984; Page, Kimsey, and Laidlaw, 1984; Moritz, 1983, 1985).

Factors Determining Multiple Mating

The mechanics of insemination and sperm mixing are difficult to understand, because most sperm that enter the queen are passed back out the sting chamber. The elimination of excess sperm does ensure that many drones are represented in the queen's offspring, since a small portion of each drone's ejaculate is taken into the limited volume of the spermatheca. The utilization of sperm from multiple males has far-reaching consequences for colony structure; it creates groups of sisters in the colony which might be expected to work for their own survival and reproduction, rather than that of their half-sisters, which have a different drone father.

The explanation for the evolution of multiple mating in honey bees may lie in their sex determination system and its relationship to brood viability. After fertilization, sex is determined at a single locus with multiple alleles; estimates of the number of alleles at this locus range from 6 to 18 (Mackensen, 1955; Laidlaw, Gomes, and Kerr, 1956; Kerr, 1967; Woyke, 1976a; Adams et al., 1977). Those individuals heterozygous at this locus develop into normal females, whereas homozygosity results in diploid males (Mackensen, 1951; see also Chapter 4). These diploid males are eaten by workers as young larvae, because they are not competitive with normal haploid males which result from unfertilized eggs. If a queen mates with a male having one sex allele identical to hers, the mortality rate of her fertilized offspring will be 50%, since half of her eggs will be homozygous at the sex-determining locus. Thus, any mechanism to increase heterozygosity at this locus will be strongly favored by selection.

One such mechanism may be multiple mating (Page, 1980). The likelihood of homozygosity at the sex-determining locus decreases dramatically with increased matings, and this function appears to reach its asymptote at about ten matings, close to the average number of drones which mate honey bee queens. By multiple mating, queens are decreasing brood mortality and increasing variability in their offspring, thereby minimizing the potential impact of mating with an undesirable drone.

The implications of multiple mating extend far beyond improving brood viability; it results in patrilineal sister groups within colonies. Coefficients of relatedness can be calculated which indicate the average proportion of their genotypes held in common, and these relationships are strongly influenced by the fact that males are haploid and females diploid. The effect of haplo-diploidy on relatedness is that sisters show a coefficient of relatedness of 0.75, whereas for half-sisters the coefficient is only 0.25. Thus, the activities of workers should be directed toward improving the fitness of their own sisters rather than that of their half-sisters.

The functioning of patrilineal kinship subgroups within colonies requires that workers be able to recognize their sisters, and there is a growing body of evidence that kin recognition occurs between workers and both larvae and adults in the nest (for reviews, see Breed,

1985). Research into this topic has used the technique of artificially inseminating queens with drones that produce offspring of particular shades and patterns so that the degree of relatedness between individuals can be determined. Testing for worker-worker recognition has primarily involved laboratory tests in which groups of sisters are maintained in isolation for a period of time, then either sisters, half-sisters, or unrelated workers are introduced into a test arena, and the level of aggressive interactions between individuals is recorded. Sisters are accepted into these groups much more readily than either unrelated individuals or half-sisters, indicating that kin recognition between adult workers can occur (Breed, 1983b; Getz and Smith, 1983). The recognition factor is evidently olfactory, since workers can be conditioned to discriminate between odors from full and half-sisters (Getz, unpublished observations).

Similar observations have been made of recognition between workers and both larval and adult queens. Swarming workers can discriminate between their own and a foreign queen, and even between a sister of their own queen and an unrelated queen (Boch and Morse, 1974, 1979, 1981, 1982; Ambrose, Morse, and Boch, 1979). In arena experiments similar to those used to test for worker-worker kin recognition, queens have been removed from small groups of attendant workers and replaced with sister or unrelated queens, and the workers have become more aggressive toward the unrelated queens (Breed, 1981). Thus, some recognition of degree of relatedness occurs between workers and adult queens, although preliminary reports of workers differentiating between sister and half-sister queens during swarming have not been substantiated (Getz, Bruckner, and Parisian, 1982; Getz and Winston, unpublished observations).

Patrilines may also influence which larvae are reared as queens by workers. If kin recognition functions in colonies, then sister workers should preferentially rear larvae of their own sisters rather than of their half-sisters. When workers are given large numbers of queen cells containing larvae to which they are related by a coefficient of 0.75, 0.31, or 0.25, workers preferentially rear the closely related larvae over the distantly related individuals (Page and Erickson, 1984). Subsequent experiments have confirmed these results and also shown that workers use cues of genetic rather than environmental origin to determine the degree of relatedness between themselves and larvae (Visscher, 1986). However, in experiments in which workers have been given choices of sister versus unrelated larvae, no such discrimination in queen rearing has been found (Breed, Velthuis, and Robinson, 1984). This may not be a surprising result—workers would rarely be in a natural situation in which they were confronted with totally unrelated individuals to rear. But commercial queen producers frequently rear large numbers of queens in colonies containing workers to which the queens are not related, and the lack of a rejection response between workers and unrelated queen larvae explains the success of this management technique. Also, larvae in these experiments were transferred to cells containing royal jelly, which may mask kin-related odors (Visscher, 1986).

Although the potential for kin recognition within colonies is becoming increasingly clear, the functioning of this capacity has not yet been demonstrated. If workers promote the fitness of the individuals to which they are most closely related, their full sisters, then preferences for larval feeding of both workers and queens, selection of more closely related virgin queens during colony requeening episodes, and reduced levels of aggressive behavior between kinship subgroups within colonies should all be demonstrable. Future research into kin recognition will undoubtedly move beyond the laboratory and theoretical phase to examine how patrilineal subgroups based on kinship function in active colonies.

13

The Biology of Temperate and Tropical Honey Bees

One of the most striking aspects of honey bee biology is the variability found within and between races of *Apis mellifera*. Bees vary in behavioral, morphological, and physiological characteristics such as color, size, tongue length, defensive behavior, nesting biology, dialects of dance language, and susceptibility to diseases. This natural variability is not surprising, considering that the many honey bee races have evolved in response to the diverse ecological conditions found in the temperate and tropical regions of Eurasia and Africa. For centuries beekeepers have taken advantage of this variability to select for economically desirable characteristics, and bees of many races have been imported to new regions of the world as beekeeping developed. This large-scale movement of bees has provided an opportunity to investigate the characteristics of temperate- and tropical-evolved bees in new environments as well as the habitats in which they evolved.

Many of the characteristics exhibited by different bee races express the influence of seasonal factors, including climate and resource abundance. In temperate and subtropical climates winter conditions result in dramatic changes in honey bee biology. Brood rearing and foraging are severely curtailed or stopped, adult longevity rises, and workers and the queen become quiescent, forming a winter cluster which obtains energy to generate heat from stored honey. For wet tropical areas, however, "winter" has a very different meaning, since temperature differences between seasons are not great, and it is rainfall which determines seasonality. For bees the season is most strongly reflected in the timing and extent of flowering and nectar and pollen production. In Africa, for example, the dry season is the dearth period when little bee forage is available, whereas in most areas of South America the wet season is the period of diminished flowering. Whatever the season when flowering is reduced, there are differences in honey bee biology based on these seasonal factors, differences as profound as those between summer and winter in colder climates.

Climate and patterns of resource abundance are not the only factors which have influenced the evolution of differences between temperate and tropical bee races; predation has also been a strong selective force. Although a direct comparison of predation pressure on honey bees in temperate and tropical habitats has not been made, the overwhelming number of predatory species and individuals in tropical habitats, as compared to those in temperate regions, clearly results in higher predation pressure in tropical habitats (see Chapter 7). One effect of high predation rates is that tropical-evolved bees are considerably more aggressive than their temperate counterparts, but other aspects of bee biology have also been strongly influenced by predation pressure and are discussed in this chapter. (For recent reviews of the topics that follow, see Winston, Taylor, and Otis, 1983, and Seeley, 1985a.)

Seasonal Patterns of Colony Demography

Bees in temperate climates respond more strongly to seasonal factors than do tropical bees, as is evidenced by their patterns of colony demography, particularly seasonal patterns of brood rearing and worker longevity. Colonies in temperate areas show rapid increases in brood rearing during the spring, peaking in May and June, with a gradual decline in brood rearing until the fall; little or no brood is reared during most of the winter (Nolan, 1925, 1928; Gontarski, 1953; Fukuda and Sekiguchi, 1966). A similar pattern has been observed in subtropical areas of Israel and Egypt for European and Cyprian races, with brood rearing reduced or absent from November to February (Bodenheimer, 1937; Bodenheimer and Ben-Nerya, 1937; Hassanein and El-Banby, 1960). In contrast, brood rearing is continuous throughout the year in wet tropical habitats, although at a reduced rate at the height of the dearth season (Winston, 1980).

Colonies of temperate-evolved bees also show more pronounced seasonal differences in worker longevity than do tropical-evolved bee races. Worker life spans of 150 days or longer are common in overwintering colonies in cold temperate areas (Fukuda and Sekiguchi, 1966), whereas workers in tropical situations show average life spans of only 20–25 days under dearth conditions, in contrast to the mean longevities of 12–18 days characteristic during the flowering period (Winston, 1979c, 1980; Winston and Katz, 1982). The difference between dearth and abundance periods is obviously much more dramatic under temperate conditions. The slightly prolonged longevity characteristic of tropical workers during the dearth season is probably due to decreased foraging and brood rearing during that time, as reported for other races (Maurizio, 1950; Fukuda and Sekiguchi, 1966;

Sakagami and Fukuda, 1968), while the more pronounced differences found in temperate climates can be attributed to more dramatic changes in overwintering physiology (see Chapter 7). Although the extent of seasonal differences in longevity is partly determined by environment, it also has a genetic component, since workers in colonies of temperate-evolved bees maintained under tropical conditions exhibit somewhat longer life spans than similarly maintained workers of tropical races, and tropical-evolved workers have much shorter life spans under temperate conditions than do temperate-evolved bees (Woyke, 1973a; Winston and Katz, 1982).

Swarming and Reproductive Biology

Possibly the most dramatic differences between temperate and tropical bees is their reproductive biology, particularly overall reproductive rates and relative investment in drone production. The swarming rate of tropical-evolved honey bees is particularly impressive; Otis (1980) observed that a single colony of Africanized bees in French Guiana produces between 6 and 12 swarms a year, and, when the swarm production of these offspring is included, almost 60 colonies can result from the single starting colony after only 1 year. This high rate of increase results from colonies swarming 3 or 4 times during the 8-month swarming season, often with intervals as short as 50 days between prime swarms, and the production of an average of almost 2 afterswarms in addition to the prime swarm during each swarming episode. The actual growth rate of feral colonies in French Guiana is approximately 16-fold annually, despite high losses due to absconding and colony death. Similar high swarming rates are characteristic of tropical bees in Africa (Chandler, 1976; Fletcher, 1977a, 1978a; Fletcher and Tribe, 1977b; Dutton et al., 1981).

In contrast, the highest reported swarming rate for unmanaged European bees during a favorable temperate growing season is only 3.6 swarms annually, and other studies have shown swarm production of only 1.0–2.6 annually (Winston, 1980; Lee, 1985; Morales, 1986). Annual population growth estimates for temperate-evolved bees range from 0 to three times, much lower than the 16-fold increase reported for tropical-evolved bees in South America (Seeley, 1978; Winston, 1980; Winston, Taylor, and Otis, 1983). This difference in swarming rate evidently has a strong genetic component, since only 2 of 17 unmanaged European bee colonies swarmed during the 4 months most favorable for swarming in French Guiana (Winston, 1980). These results are consistent with observations of beekeepers in tropical Africa and South America, who frequently report little swarming with European bees but excessive swarming with Africanized bees in the same area.

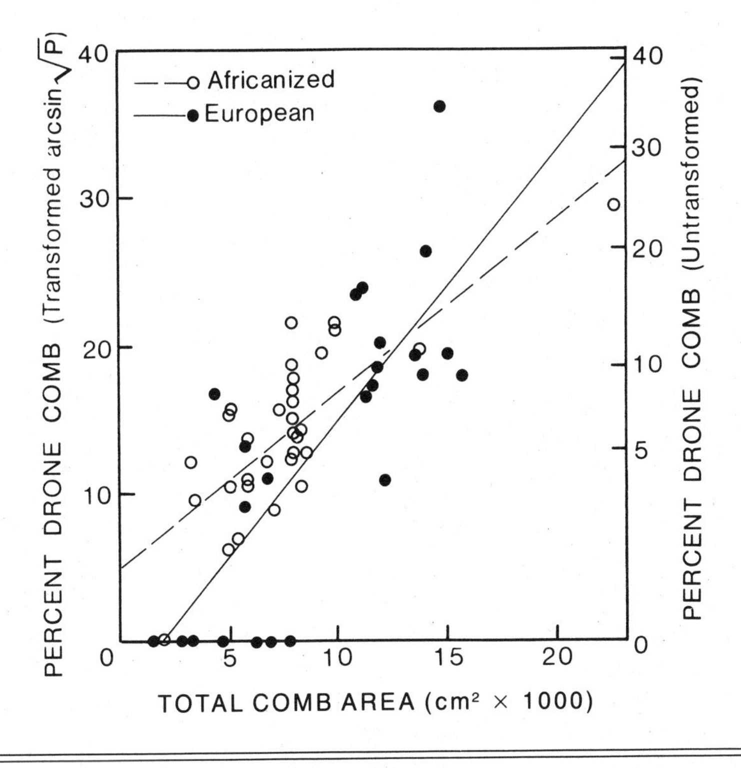

Fig. 13.1

The relationship between the percentage of drone comb and the total amount of comb constructed by colonies of tropical-evolved Africanized bees in South America (data from Otis, 1980) and temperate-evolved European bees in North America (data from Lee and Winston, 1985b). Data were plotted and analyzed on the transformed scale shown on the left axis; for comparison the untransformed scale is shown on the right. (Redrawn from Lee and Winston, 1985b.)

Temperate- and tropical-evolved honey bee races also differ in the percentage of drone comb constructed at similar colony sizes (Otis, 1980; Lee and Winston, 1985b). Tropical Africanized honey bees in South America construct a higher percentage of drone comb at small colony sizes than do temperate European races (Fig. 13.1). Also, small colonies of temperate-evolved bees do not initially allocate as many resources to drone comb production, but rather delay the construction of drone comb in favor of comb with worker-sized cells.

Absconding

In the tropics colonies of African and Africanized bees abscond much more often than European-evolved colonies maintained in either temperate or tropical habitats. Absconding can be defined as the abandoning of a nest by a colony which forms a swarm and presumably reestablishes itself elsewhere. Absconding swarms differ from reproductive swarms in that few or no workers and no adult or viable immature queens are left behind in the original colony. Absconding from feral nests is generally either disturbance induced or resource induced.

Disturbance-induced absconding is rare in temperate habitats, partly because there are few predators to induce such absconding, but also because temperate-evolved bees have a much lower tendency to abandon their nest. Among tropical-evolved bees, absconding due to disturbance usually results from partial or total destruction of colonies by predators, destruction of comb by wax moths, fire in the proximity of the nest, heavy wasp or bird predation at the nest entrance, inability to regulate temperature due to cold or to excessive sunlight, and rain entering the nest (Fletcher, 1975/1976, 1978a; Chandler, 1976; Woyke, 1976b; Winston, Otis, and Taylor, 1979). In these cases absconding generally occurs within hours of, or at most a few days after, the disturbance.

Resource-induced absconding seems to result from a scarcity of nectar, pollen, or water and occurs primarily during the dearth season in tropical habitats. Tropical African and Africanized colonies frequently abscond during these dearth periods; in French Guiana, 30% of Africanized colonies abscond during the wet season, when there is relatively little flowering, presumably to search for areas with better resources (Winston, Otis, and Taylor, 1979). Similarly, 79% of absconding swarms recorded in Brazil by Cosenza (1972) occurred during a resource dearth period. Absconding rates of tropical African bees are generally 15–30% a year, and can be as high as 100% in some areas. The highest incidence of African absconding occurs during the dry season, when there is less flowering, and also less water at a time when bees need it for regulating internal nest temperature (Smith, 1960; Fletcher, 1975/1976, 1978a; Woyke, 1976b). In contrast, resource-induced absconding by European bees in either temperate or tropical regions is infrequently recorded (Martin, 1963; Winston, Otis, and Taylor, 1979; Robinson, 1982). Under wet season conditions in French Guiana, colonies of European bees maintained in the same manner as Africanized bees all dwindled and died instead of absconding, and interviews with African and South American beekeepers indicate that absconding among European bees is also infrequent under managed conditions (Winston, Otis, and Taylor, 1979).

Resource-induced absconding differs from abscondings due to disturbance in the pattern of colony preparation prior to absconding. Whereas disturbance can induce a colony to leave within hours or days, careful preparations are made by colonies for many weeks prior

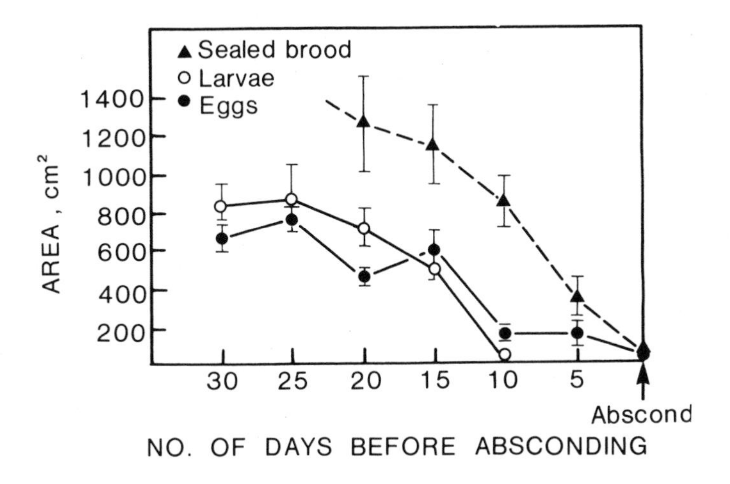

Fig. 13.2

Areas of eggs, larvae, and sealed brood in colonies of Africanized bees prior to absconding. (Redrawn from Winston, Otis, and Taylor, 1979.)

to abandoning the nest due to a resource shortage (Winston, Otis, and Taylor, 1979). Colonies preparing to abscond begin reducing their brood rearing about 25 days before absconding and rear no new larvae in the 10–15 days preceding absconding (Fig. 13.2). The queen continues to lay a few eggs up to the abscond date, although these are evidently consumed by the workers rather than reared. Most of the stored pollen is also consumed prior to absconding, as is much of the stored honey. Interestingly, this type of seasonal absconding is timed so that the absconding swarm does not issue until the last sealed brood has emerged; in this way absconding swarms have a reasonable population of young workers with which to initiate the new nest. Also, workers engorge with honey before absconding, so that the absconding colony only leaves their wax comb behind, taking all of its other resources with it in the form of newly emerged workers, honey, and protein from the recently consumed pollen, which is probably stored in the workers' fat bodies and hypopharyngeal glands.

Deteriorating resource conditions are not sufficient to explain absconding, however, since many colonies maintained at the same sites as absconding colonies persist throughout the dearth season. Resource-induced absconding is most likely induced by a combination of poor forage and internal colony conditions, particularly chronic low brood survival (Winston, Otis, and Taylor, 1979; Winston, 1980). It is interesting that there are no differences in the amounts of honey, pollen, or brood areas between absconding and persisting colonies, but there are differences in the timing of swarming relative to the absconding season. Colonies that abscond are generally those that have swarmed within 6 weeks of the onset of a dearth period, which results in their having lower worker populations, more older workers, and higher brood mortality than colonies that swarmed more than 6 weeks before the absconding season.

Whatever the cause of absconding, colonies which abandon their nests may migrate long distances in search of better foraging conditions. Workers in absconding swarms carry almost twice the amount of honey as workers in reproductive swarms, providing fuel for longer dispersal distances (Otis, Winston, and Taylor, 1981). The behavior of workers in absconding swarms also differs from reproductive swarms in that they frequently do not scout for new nest sites upon emergence but travel cross-country for undetermined distances before scouting. In addition, absconding swarms settle at interim sites and send out workers to search for floral resources, possibly scouting new areas for resource quality before searching for a nest site. Absconding swarms may travel as far as 160 km or more before constructing a new nest, migrating through areas of poor resources until they discover a better area with abundant but localized floral resources, such as Eucalyptus plantations in Africa (Nightingale, 1976; Fletcher, 1978a).

Other Differences

Other differences between temperate- and tropical-evolved honey bees have been reviewed in previous chapters but may be briefly summarized here. First, the nests of tropical honey bee races are about one-third smaller and store considerably less honey than do temperate races, and also are frequently constructed outside of cavities (Chapter 5). Second, individual workers in tropical-evolved colonies have shorter development periods and adult life spans, begin foraging at younger ages, and are smaller than workers from temperate regions (Chapter 4). Finally, tropical bee races are considerably more aggressive than temperate races (Chapter 7).

Factors Selecting for Temperate/Tropical Differences

The differences between the two types of races are most pronounced when it comes to factors important for survival under feral, unmanaged conditions (Table 13.1). Many of these characteristics are interrelated; for example, the high swarming rates of tropical bees result in part from their smaller nest size and amount of honey storage, so that more energy goes into reproduction than nest construction. The

Table 13.1 Differences between temperate- and tropical-evolved honey bees

	Environment	
Feature	Temperate	Tropical
Nest architecture		
Colony size	Large	Small
Honey storage	Large	Small
Exposed nests	Rare	Common
Swarming rate	Low	High
Absconding	Rare	Common
Colony defense	Moderate	Intense
Worker characteristics		
Development period	Long	Short
Adult life span	Long	Short
Foraging age	Old	Young
Body size	Large	Small

Source: Winston, Taylor, and Otis, 1983.

smaller worker and cell size, shorter duration of development and life span, and earlier foraging age that characterize tropical bees are all related traits which result in accelerated colony growth and hence increased swarming rates.

The differences between bees which evolved in these diverse habitats seem to reflect racial differences that are not related solely to environment. Studies of tropical African bees of various races in Africa have shown characteristics similar to those described for Africanized bees in South America. European colonies in tropical South America are generally similar to those in temperate North America, at least in such characteristics as swarming and absconding rates and worker development periods, although worker life span is shorter in tropical than in temperate regions. Thus, differences between the temperate- and tropical-evolved groups have genetic components as well as environmentally influenced characteristics.

These differences can be partly explained by considering the contrasting selective pressures in temperate and tropical habitats which might result in different individual worker and colony characteristics. The most obvious such factor is climate. Bees in temperate areas maintain the temperature of the winter cluster by consuming stored honey. Winter survival in these regions requires substantial honey reserves, and therefore a relatively large nest, inside a well-insulated cavity. Tropical bees, in contrast, do not experience a prolonged cold season and do not require large nests; their nests are smaller and contain less honey, and the colony can survive outside a cavity.

The timing and extent of swarming is also partly determined by climatic factors. In order to increase the chances of survival, a temperate-evolved colony should swarm early enough in the season to have sufficient time for the swarm and the original colony to store enough honey for the next winter. Multiple swarming episodes each year could leave colonies with insufficient honey to support them throughout the winter. In the wet tropics bees are free of these cold temperature restrictions and can commit resources to swarm production rather than to honey storage.

The tendency to abscond is partly determined by climatic factors and their effects on flowering and nectar secretion. Honey bees have two strategies for surviving a dearth season, hoarding and absconding. Under temperate conditions hoarding is the only possible strategy, since absconding swarms cannot collect sufficient nectar to survive the next winter. As a consequence, temperate bees rarely abscond. Under tropical conditions, however, either absconding or hoarding could be an effective strategy. It might be advantageous for colonies to abscond if resources were patchy or nonexistent, or if the duration of the dearth season was unpredictable. In tropical areas many colonies abscond and presumably search for better forage rather than utilize stored honey to survive a dearth season.

Predation is a second factor which may be important in determining colony characteristics, at least under tropical conditions; in temperate areas predation is a relatively insignificant cause of colony mortality. The extreme aggressive behavior often exhibited by tropical bees is undoubtedly a defense against predators, which are nevertheless successful in killing many colonies. The short intervals between successive swarming, and the large number of swarms produced, have the effect of offsetting these high colony death rates. Moreover, the early drone production characteristic of newly founded colonies of tropical bees improves the probability of some reproduction should predation occur prior to swarming. In addition, absconding enables an attacked colony to reestablish itself elsewhere. In contrast, temperate colonies in the tropics are more likely to be attacked before swarming than colonies of tropical bees, both because of the longer period before they attain sufficient size to swarm and because of the attractiveness of a large nest to predators.

The characteristics described here may explain some notable successes and failures of feral honey bee populations in becoming established after importations. Given the temperate-evolved qualities of European bees, it is not surprising that their introductions to tropical countries have not met with success. Feral colonies of European bees have not become established in South and East Africa despite many attempted importations, nor have they become widely established in tropical South America (Fletcher, 1977b, 1978a; Winston, 1979d). Surprisingly, imported European bees did establish a feral population in New Guinea, although its range expanded at only 14 km/year (Michener, 1963a,b). Importations of European bees to subtropical and temperate areas have had notable success, as evidenced by the large number of feral colonies and extensive beekeeping from Mexico to Canada, and in Australia, Japan, and the People's Republic of China. Feral Africanized bees, however, have developed large populations and spread extensively through most of South America. Their natural spread has slowed or stopped as they have encountered the more temperate areas of Uruguay and Argentina, suggesting that the characteristics which have enabled them to do so well in the tropics are not necessarily suitable for survival in temperate climates (Taylor, 1977; Kerr, Rio, and Barrionuevo, 1982; Dietz, Krell, and Eischen, 1985).

There is a third selective pressure on bee colonies, in addition to climate and predation, which has probably shaped the characteristics of temperate and tropical honey bees: the interaction between resource abundance and worker and colony foraging traits. A mounting body of evidence suggests that fundamental differences in foraging behavior between temperate- and tropical-evolved honey bees can be explained by differences in the qualities and characteristics of honeyflows in the two habitats. Tropical flowers tend to produce more dilute nectar and be spatially distributed in a more patchy fashion than in temperate habitats, where large areas containing concentrated numbers of high nectar-yielding plants are common. These differences in resource characteristics appear to have selected for various differences in foraging behavior between bees evolving in their respective habitats.

Generally, tropical bees tend to do better under weak honeyflows than do temperate bees. In one comparison made during two periods of different nectar availability in Venezuela, European bees collected more nectar per day than Africanized bees during the more productive dry season, but the Africanized bees were superior under wet season conditions of low nectar availability (Rinderer, Collins, and Tucker, 1985). In another study Africanized bees stored between two and four times the amount of nectar that European colonies stored during one honeyflow period in Brazil (Portugal-Araujo, 1971). Also, under wet season conditions the Africanized bees either maintain themselves or collect sufficient nectar to abscond, whereas European bees dwindle and die (Winston, Otis, and Taylor, 1979; Winston, 1980).

The foraging behavior of individual workers and the recruitment behavior of colonies explains in part why tropical bees collect more nectar under weak flow conditions than do temperate bees (Nuñez, 1979; Winston and Katz, 1982; Rinderer et al., 1984; Rinderer, Collins, and Tucker, 1985). Basically, tropical-evolved bees forage more as individuals while temperate-evolved bees are superior recruiters. Comparisons made in South America have shown that Africanized bees have shorter inspection times at feeders, switch from a declining resource more quickly, make more trips per hour, and prefer higher nectar concentrations than do European bees. At the colony level Africanized bees dance less and thus show reduced recruitment relative to European bees, which devote a higher proportion of their worker population to foraging than do Africanized bees and thus require stronger honeyflows to balance energy gains and losses.

But the key experiments linking honeyflow conditions to foraging behavior have yet to be done, partly because it is extraordinarily difficult to devise an experimental design that allows colonies to forage normally and simultaneously permits the precise measurement of nectar availability in the field. The importance of pollen, too, in determining foraging behavior has not been compared between bees of various races, and indeed is not clearly understood for any race of honey bee. Research into these topics is fundamental to our understanding of honey bee biology because food collection is so important in determining almost all other aspects of bee behavior, from swarming to overwintering and absconding.

References

Author Index

Subject Index

References

Abdellatif, M. A. 1965. Comb cell size and its effect on the body weight of the worker bee, *Apis mellifera* L. *Amer. Bee J.* 105:86–87.

Adam, A. 1912. Bau und Mechanismus des Receptaculum seminis bei den Bienen, Wespen, und Ameisen. *Zool. Jb. Anat. Ontog. Tiere* 35:1–74.

Adams, J., E. D. Rothman, W. E. Kerr, and Z. L. Paulino. 1977. Estimation of the number of sex alleles and queen matings from diploid male frequencies in a population of *Apis mellifera. Genetics* 86:583–596.

Alber, M. 1956. The size of comb cells as a racial characteristic. Paper presented at Sixteenth International Beekeeping Congress, Vienna.

Alfonsus, E. C. 1932. Swarming and supercedure. *Wis. Beekeeping* 8:34–36.

——— 1933. Zum Pollenverbrauch des Bienenvolkes. *Arch. Bienenk.* 14:220–223.

Alford, D. V. 1975. *Bumblebees.* London, Davis-Poynter.

Allen, M. D. 1955. Observations of honeybees attending their queen. *Brit. J. Anim. Behav.* 3:66–69.

——— 1956. The behaviour of honeybees preparing to swarm. *Brit. J. Anim. Behav.* 4:14–22.

——— 1957. Observations on honeybees examining and licking their queen. *Brit. J. Anim. Behav.* 5:81–84.

——— 1959a. The occurrence and possible significance of the "shaking" of honeybee queens by the workers. *Anim. Behav.* 7:66–69.

——— 1959b. The "shaking" of worker honeybees by other workers. *Anim. Behav.* 7:233–240.

——— 1960. The honeybee queen and her attendants. *Anim. Behav.* 8:201–208.

——— 1963. Drone production in honeybee colonies. *Nature* 199:789–790.

——— 1965a. The production of queen cups and queen cells in relation to the general development of honeybee colonies and its connection with swarming and supersedure. *J. Apic. Res.* 4:121–141.

——— 1965b. The effect of a plentiful supply of drone comb on colonies of honeybees. *J. Apic. Res.* 4:109–119.

Allen, M. D., and E. P. Jeffree. 1956. The influence of stored pollen and of colony size on the brood rearing of honeybees. *Ann. Appl. Biol.* 44:649–656.

Altenkirch, G. 1962. Untersuchungen über die Morphologie der abdominalen Hautdrusen einheimischer. *Apiden. Zool. Beitr.* (Berlin) 7:161–238.

Al-Tikrity, W. S., A. W. Benton, R. C. Hillman, and W. W. Clarke, Jr. 1972. The relationship between the amount of unsealed brood in honeybee colonies and their pollen collection. *J. Apic. Res.* 11:9–12.

Ambrose, J. T. 1976. Swarms in transit. *Bee World* 57:101–109.

——— 1978. Birds. In *Honey bee pests, predators, and diseases,* ed. R. A. Morse, pp. 215–226. Ithaca, Cornell Univ. Press.

Ambrose, J. T., R. A. Morse, and R. Boch. 1979. Queen discrimination by honey bee swarms. *Ann. Entomol. Soc. Amer.* 72:673–675.

Anderson, J. 1931. How long does a bee live? *Bee World* 12:25–26.

Anderson, L. M., and A. Dietz. 1976. Pyridoxine requirement of the honey bee (*Apis mellifera*) for brood rearing. *Apidologie* 7:67–84.

Anderson, R. H. 1963. The laying worker in the Cape honey bee, *Apis mellifera capensis*. *J. Apic. Res.* 2:85–92.

Arnhart, L. 1923. Das krallenglied der honigbiene. *Arch. Bienenk.* 5:37–86.

Arnold, G., and B. Delage-Darchen. 1978. Nouvelles données sur l'équipment enzymatique des glandes salivaires de l'ouvriere d'*Apis mellifera* (Hyménoptère Apidé). *Annales des Sciences Naturelles, Zool. Biol. Anim.* 20:401–422.

Asencot, M., and Y. Lensky. 1976. The effect of sugars and juvenile hormone on the differentiation of the female honey bee (*Apis mellifera* L.) larvae to queens. *Life Sci.* 18:693–700.

——— 1984. Juvenile hormone induction of "queenliness" on female honey bee (*Apis mellifera* L.) larvae reared on worker jelly and on stored royal jelly. *Comp. Biochem. Physiol.* 78b:109–117.

Autrum, H. J., and M. Stoecker. 1950. Die Verschmelzungsfrequenzen des Bienenauges. *Naturforsch.* 5b:38–43.

Avitabile, A., and J. P. Kasinskas. 1977. The drone population of natural honeybee swarms. *J. Apic. Res.* 16:145–149.

Avitabile, A., R. A. Morse, and R. Boch. 1975. Swarming honey bees guided by pheromones. *Ann. Entomol. Soc. Amer.* 68:1079–82.

Avitabile, A., D. Stafstrom, and K. J. Donovan. 1978. Natural nest sites of honeybee colonies in trees in Connecticut, USA. *J. Apis Res.* 17:222–226.

Back, E. 1956. Einfluss der im Pollen enthaltenen Vitamine auf Lebensdauer, Ausbildung der Pharynxdrüsen and Brutfähigkeit der Honigbiene. *Insectes Sociaux* 3:285–292.

Bailey, L. 1952. The action of the proventriculus of the worker honeybee. *J. Exp. Biol.* 29:310–327.

——— 1954. The respiratory currents in the tracheal system of the adult honeybee. *J. Exp. Biol.* 31:589–593.

Barbier, M., and D. Bogdanovsky. 1961. Isolement et identification du méthylène-24 cholestérol, à partir des larves de reines d'abeilles, et de la gelée royale. *Compt. Rend. Acad. Sci.* (Paris) 252:3497–98.

Barbier, M., and M. F. Hugel. 1961. Synthèse de l'acide cèto-9-décèn-2-ciso.que, isomère de la "Substance Royale." *Bull. Soc. Chim. Fr.* 202:1324–26.

Barbier, M., and E. Lederer. 1960. Structure chemique de la "substance royale" de la reine d'abeille (*Apis mellifica*). *Compt. Rend. Acad. Sci.* (Paris) 250:4467–69.

Barbier, M., E. Lederer, and T. Nomura. 1960. Synthèse de l'acide cèto-9-décène-2-trans-oïque (substance royale) et de l'acide cèto-8-nonène-2-trans-oïque. *Compt. Rend. Acad. Sci.* (Paris) 251:1133–35.

Barker, R. J. 1971. The influence of food inside the hive on pollen collection by a honeybee colony. *J. Apic. Res.* 10:23–26.

Barker, R. J., and Y. Lehner. 1972. The resistance of pollen grains and their degradation by bees. *Bee World* 53:173–177.

Barker, S. A., A. B. Foster, D. C. Lamb, and N. Hodgson. 1959. Identification of 10-hydroxy-2-decenoic acid in royal jelly. *Nature* 183:996–997.

Bastian, J., and H. Esch. 1970. The nervous control of the indirect flight muscles of the honey bee. *Z. vergl. Physiol.* 67:307–324.

Becker, F. 1925. Die Ausbildung des Geschlechtes bei der Honigbiene, II. *Erlanger Jb. Bienenk.* 3:163–223.

Beetsma, J. 1979. The process of queen-worker differentiation in the honeybee. *Bee World* 60:24–39.

Belik, M. Y. 1979. On the sub-microscopic organization of the cells of the Nasonov gland. *Pchelovodstvo* 1:16–18 (in Russian).

Betts, A. D. 1935. The constancy of the pollen-collecting bee. *Bee World* 16:111–113.

Bisetzky, A. A. 1957. Die Tänze der Bienen nach einem Fussweg zum Futterplatz. *Z. vergl. Physiol.* 40:264–288.

Bishop, G. H. 1920a. Fertilization in the honey-bee, I. The male sexual organs: their histological structure and physiological functioning. *J. Exp. Zool.* 31:225–265.

——— 1920b. Fertilization in the honey-bee, II. Disposal of the sexual fluids in the organs of the female. *J. Exp. Zool.* 31:267–286.

Blum, M. S., H. M. Fales, K. W. Tucker, and A. M. Collins. 1978. Chemistry of the sting apparatus of the worker honeybee. *J. Apic. Res.* 17:218–221.

Blum, M. S., R. Boch, R. E. Doolittle, M. T. Tribble, and J. G. Traynham. 1971. Honey bee sex attractant: conformation analysis, structural specificity and lack of masking activity of congeners. *J. Insect Physiol.* 17:349–364.

Boch, R. 1979. Queen substance pheromone produced by immature queen honeybees. *J. Apic. Res.* 18:12–15.

Boch, R., and Y. Lensky. 1976. Pheromonal control of queen rearing in honeybee colonies. *J. Apic. Res.* 15:59–62.

Boch, R., and R. A. Morse. 1974. Discrimination of familiar and foreign queens by honeybee swarms. *Ann. Entomol. Soc. Amer.* 67:709–711.

——— 1979. Individual recognition of queens by honeybee swarms. *Ann. Entomol. Soc. Amer.* 72:51–53.

——— 1981. Effects of artificial odors and pheromones on queen discrimination by honey bees (*Apis mellifera* L.) *Ann. Entomol. Soc. Amer.* 74:66–67.

——— 1982. Genetic factor in queen recognition odors of honey bees. *Ann. Entomol. Soc. Amer.* 75:654–656.

Boch, R., and W. C. Rothenbuhler. 1974. Defensive behaviour and production of alarm pheromone in honeybees. *J. Apic. Res.* 13:217–221.

Boch, R., and D. A. Shearer. 1962. Identification of geraniol as the active component in Nassanoff pheromone of the honey bee. *Nature* 194:704–706.

——— 1964. Identification of nerolic and geranic acids in the Nassanoff pheromone of the honey bee. *Nature Lond.* 202:320–321.

——— 1966. Iso-pentyl acetate in stings of honeybees of different ages. *J. Apic. Res.* 5:65–70.

Boch, R., D. A. Shearer, and A. Petrasovits. 1970. Efficacies of two alarm substances of the honey bee. *J. Insect Physiol.* 16:17–24.

Boch, R., D. A. Shearer, and R. W. Shuel. 1979. Octanoic and other volatile acids in the mandibular glands of the honeybee and in royal jelly. *J. Apic. Res.* 18:250–252.

Boch, R., D. A. Shearer, and B. C. Stone. 1962. Identification of iso-amyl acetate as an active component in the sting pheromone of the honey bee. *Nature* 195:1018–20.

Boch, R., D. A. Shearer, and J. C. Young. 1975. Honey bee pheromones: field tests of natural and artificial queen substance. *J. Chem. Ecol.* 1:133–148.

Bodenheimer, F. S. 1937. Studies in animal populations, II. Seasonal population trends of the honey-bee. *Q. Rev. Biol.* 12:406–425.

Bodenheimer, F. S., and A. Ben Nerya. 1937. One year's studies on the biology of the honey-bee in Palestine. *Ann. Appl. Biol.* 24:385–403.

Boehm, B. 1965. Relations between fat-bodies, oenocytes and wax-gland development in *Apis mellifica* L. *Z. Zellforsch.* 65:74–115.

Bottcher, F. K. 1975. Beiträge zur Kenntnis des Paarungsfluges der Honigbiene. *Apidologie* 6:233–281.

Bozina, K. D. 1961. How long does the queen live? *Pchelovodstvo* 38:13 (in Russian).

Breed, M. D. 1981. Individual recognition and learning of queen odors by worker honeybees. *Proc. Nat. Acad. Sci.* (U.S.) 78:2635–37.

——— 1983a. Correlations between aggressiveness and corpora allata volume, social isolation, age and dietary protein in worker honeybees. *Insectes Sociaux* 30:482–495.

——— 1983b. Nestmate recognition in honey bees. *Anim. Behav.* 31:86–91.

——— 1985. How honeybees recognize their nestmates: a re-evaluation from new evidence. *Bee World* 66:113–118.

Breed, M. D., H. H. W. Velthuis, and G. E. Robinson. 1984. Do worker honey bees discriminate among unrelated and related larval phenotypes? *Ann. Entomol. Soc. Amer.* 77:737–739.

Bresslau, E. 1905. Der Samenblasengang der Bienenkönigen. *Zool. Anz.* 29:299–323.

Bridges, A. R. 1977. Fine structure of the honey bee (*Apis mellifera* L.) venom gland and reservoir: a system for the secretion and storage of a naturally produced toxin. *Microscop. Soc. Can.* 4:50–51.

Brouwers, E. V. M. 1984. Glucose/fructose ratio in the food of honeybee larvae during caste differentiation. *J. Apic. Res.* 23:94–101.

Burgett, D. M., and R. A. Morse. 1974. The time of natural swarming in honey bees. *Ann. Entomol. Soc. Amer.* 67:719–720.

Burrill, R. M., and A. Dietz. 1981. The response of honeybees to variations in solar radiation and temperature. *Apidologie* 12:319–328.

Butenandt, A., and H. Rembold. 1957. Über den Weiselzellenfuttersaft der Honigbiene, I. Isolierung, Konstitutionsermittlung und Vorkommen der 10-Hydroxy-2-decensäure. *Hoppe-Seyl Z.* 308:284–289.

Butler, C. 1609. *The feminine monarchie.* Oxford, Joseph Barnes.

Butler, C. G. 1940. The ages of the bees in a swarm. *Bee World* 21:9–10.

——— 1957a. The process of queen supersedure in colonies of honeybees (*Apis mellifera* L.). *Insectes Sociaux* 4:211–223.

——— 1957b. The control of ovary development in worker honeybees (*Apis mellifera*). *Experientia* 13:256–257.

——— 1959. Queen substance. *Bee World* 40:269–275.

——— 1960. The significance of queen substance in swarming and supersedure in honey-bee (*Apis mellifera* L.) colonies. *Proc. Roy. Entomol. Soc. London* (A) 35:129–132.

——— 1961. The scent of queen honey bees (*Apis mellifera*) that causes partial inhibition of queen rearing. *J. Insect Physiol.* 7:258–264.

——— 1966. Mandibular gland pheromone of worker honeybees. *Nature* 212:530.

——— 1967. Insect pheromones. *Biol. Rev.* 42:42–87.

——— 1974. *The world of the honeybee.* London, Collins.

Butler, C. G., and K. H. Calam. 1969. Pheromones of the honeybee—the secretion of the Nassanoff gland of the worker. *J. Insect Physiol.* 15:237–244.

Butler, C. G., and R. K. Callow. 1968. Pheromones of the honeybee (*Apis mellifera* L.): the "inhibitory scent" of the queen. *Proc. Roy. Entomol. Soc. London* (B) 43:62–65.

Butler, C. G., and E. M. Fairey. 1963. The role of the queen in preventing oogenesis in worker honeybees. *J. Apic. Res.* 2:14–18.

——— 1964. Pheromones of the honeybee: biological studies of the mandibular gland secretion of the queen. *J. Apic. Res.* 3:65–76.

Butler, C. G., and J. B. Free. 1952. The behaviour of worker honeybees at the hive entrance. *Behav.* 4:262–292.

Butler, C. G., and P. N. Paton. 1962. Inhibition of queen rearing by queen honey-bees (*Apis mellifera* L.) of different ages. *Proc. Roy. Entomol. Soc. London* (A) 37:114–116.

Butler, C. G., and J. Simpson. 1965. Pheromones of the honeybee (*Apis mellifera* L.): an olfactory pheromone from the Koschewnikov gland of the queen. *Scientific Studies, Univ. Libeice* (Czechoslovakia) 4:33–36.

——— 1967. Pheromones of the queen honey bee (*Apis mellifera* L.) which enable her workers to follow her when swarming. *Proc. Roy. Entomol. Soc. London* (A) 42:149–154.

Butler, C. G., R. K. Callow, and J. R. Chapman. 1964. 9-hydroydec-*trans*-2-enoic acid, a pheromone stabilizing honey bee swarms. *Nature* 201:733.

Butler, C. G., R. K. Callow, and N. C. Johnston. 1961. The isolation and synthesis of queen substance, 9-oxodec-trans-2-enoic acid, a honeybee pheromone. *Proc. Roy. Soc. London* (B) 155:417–432.

Butler, C. G., D. J. C. Fletcher, and D. Watler. 1969. Nest-entrance marking with pheromones by the honeybee *Apis mellifera* L., and a wasp, *Vespula vulgaris* L. *Anim. Behav.* 17:142–147.

Butler, C. G., E. P. Jeffree, and H. Kalmus. 1943. The behaviour of a population of honeybees on an artificial and on a natural crop. *J. Exp. Biol.* 20:65–73.

Butler, C. G., R. K. Callow, A. R. Greenway, and J. Simpson. 1974. Movement of the pheromone 9-oxodec-2-enoic acid, applied to the body surface of honeybees. *Ent. Exp. Appl.* 17:112–116.

Cahill, K., and S. Lustick. 1976. Oxygen consumption and thermoregulation in *Apis mellifera* workers and drones. *Comp. Biochem. Physiol.* 55a:355–357.

Cale, G. H., Sr., R. Banker, and J. Powers. 1946. Management for honey production, In *The hive and the honeybee,* ed. C. P. Dadant, pp. 355–412. Hamilton, Ill., Dadant.

——— 1968. Pollen gathering relationship to honey collection and egg laying in honey bees. *Amer. Bee J.* 108:8–9.

Callow, R. K. 1963. Chemical and biochemical problems of beeswax. *Bee World* 44:95–101.

Callow, R. K., and N. C. Johnston. 1960. The chemical constitution and synthesis of queen substance of honeybees (*Apis mellifera*). *Bee World* 41:152–153.

Callow, R. K., J. R. Chapman, and P. N. Paton. 1964. Pheromones of the honeybee: chemical studies of the mandibular gland secretion of the queen. *J. Apic. Res.* 3:77–89.

Callow, R. K., N. C. Johnston, and J. Simpson. 1959. 10-hydroxy-2-decenoic acid in the honeybee (*Apis mellifera*). *Experientia* 15:421.

Canetti, S. J., R. W. Shuel, and S. E. Dixon. 1964. Studies on the mode of action of royal jelly in honeybee development, IV. Development within the brain and retrocerebral complex of female honeybee larvae. *Can. J. Zool.* 42:229–233.

Carlson, C. W., and R. W. Brosemer. 1971. Comparative structural properties of insect triose phosphate dehydrogenases. *Biochem.* 10:2113–2119.

——— 1973. Amino acid compositions of cytochrome C from four hymenopteran species: evolutionary significance. *Syst. Zool.* 22:77–83.

Caron, D. M. 1978. *Marsupials and mammals.* In *Honey bee pests, predators, and diseases,* ed. R. A. Morse, pp. 227–256. Ithaca, Cornell Univ. Press.

——— 1979. Queen cup and queen cell production in honeybee colonies. *J. Apic. Res.* 18:253–256.

——— 1980. Swarm emergence date and cluster location in honeybees. *Amer. Bee J.* 119:24–25.

——— 1981. Congestion, seasonal cycle, and queen rearing as they relate to swarming in *Apis mellifera* L. *Ann. Entomol. Soc. Amer.* 74:134–137.

Caron, D. M., and C. W. Greve, 1979. Destruction of queen cells placed in queenright *Apis mellifera* colonies. *Ann. Ent. Soc. Amer.* 72:405–407.

Casteel, D. B. 1912. The manipulation of the wax scales of the honeybee. *Circ. U.S. Bur. Entomol.* 161:1–13.

Chadwick, P. C. 1931. Ventilation of the hive. *Glean. Bee Cult.* 59:356–358.

Chai, Ber Lin, and R. W. Shuel. 1970. Effects of supernumerary corpora allata and farnesol compounds on ovary development in the worker bee. *J. Apic. Res.* 9:19–27.

Chalmers, W. T. 1980. Fish meals as pollen-protein substitutes for honeybees. *Bee World* 61:89–96.

Chandler, M. T. 1976. The African honey bee *Apis mellifera adansonii:* the biological basis of its management. In *Apiculture in tropical climates,* ed. E. Crane, pp. 61–68. London, International Bee Research Association.

Chauvin, R. 1956. Les facteurs qui gouverment la ponte chez la reine des abeilles. *Insectes Sociaux* 3:499–504.

——— 1962. Sur l'épagine et sur les glandes tarsales d'Arnhart. *Insectes Sociaux* 9:1–5.

Chauvin, R., and P. Lavie. 1956. Investigations on the antibiotic substance in pollen. *Ann. Inst. Pasteur* 90:523–527.

Collins, A. M. 1979. Genetics of the response of the honeybee to an alarm chemical, isopentyl acetate. *J. Apic. Res.* 18:285–291.

——— 1980. Effect of age on the response to alarm pheromones by caged honey bees. *Ann. Entomol. Soc. Amer.* 73:307–309.

——— 1981. Effects of temperature and humidity on honeybee responses to alarm pheromones. *J. Apic. Res.* 20:13–18.

Collins, A. M., and M. S. Blum. 1982. Bioassay of compounds derived from the honeybee sting. *J. Chem. Ecol.* 8:463–470.

——— 1983. Alarm responses caused by newly identified compounds derived from the honeybee sting. *J. Chem. Ecol.* 9:57–65.

Collins, A. M., and K. J. Kubasek. 1982. Field test of honey bee (Hymenoptera: Apidae) colony defensive behavior. *Ann. Entomol. Soc. Amer.* 75:383–387.

Collins, A. M., and T. E. Rinderer. 1985. Effect of empty comb on defensive behavior of honeybees. *J. Chem. Ecol.* 11:333–338.

Collins, A. M., and W. C. Rothenbuhler. 1978. Laboratory test of the response to an alarm chemical, isopentyl acetate, by *Apis mellifera. Ann. Entomol. Soc. Amer.* 71:906–909.

Collins, A. M., T. E. Rinderer, J. R. Harbo, and A. B. Bolten. 1982. Colony defense by Africanized and European honey bees. *Science* 218:72–74.

Collins, A. M., T. E. Rinderer, K. W. Tucker, H. A. Sylvester, and J. J. Lackett. 1980. A model of honeybee defensive behavior. *J. Apic. Res.* 19:224–231.

Combs, G. F. 1972. The engorgement of swarming worker honeybees. *J. Apic. Res.* 11:121–128.

Cook, V. A. 1968. Severe clipping of queens' wings increases supersedure rate. *N.Z. Beekpr.* 30:32.

Corkins, C. L. 1930. The winter activity in the honeybee cluster. Report of the Iowa State Apiarist, pp. 44–49.

Corkins, C. L. and Gilbert, C. S. 1932. The metabolism of honeybees in winter cluster. *Bull. Wyo. Agric. Exp. Sta.* 187:1–30.

Cosenza, G. W. 1972. Estudo dos enxames de migracao de abelhas africans. Proceedings of the First Brazilian Congress of Apiculture, pp. 128–129.

Coss, R. G., and J. G. Brandon, 1982. Rapid changes in dendritic spine morphology during the honeybee's first orientation flight. In *The biology of social insects,* eds. M. D. Breed, C. D. Michener, and H. E. Evans, pp. 338–342. Boulder, Colo., Westview Press.

Costa-Leonardo, A. M. C. 1980. Morphologic studies of the secretory cycle of the mandibular glands of *Apis mellifera* (Hymenoptera, Apidae). *Rev. Brasil. Entomol.* 24:142–152.

——— 1985. Developmental cycle of the mandibular glands of *Apis mellifera* workers, II. Effect of queenlessness. *J. Apic. Res.* 24:76–79.

Crane, E. 1975. *Honey: A comprehensive survey,* ed. E. Crane. London, Heinemann.

——— 1983. *The archaeology of beekeeping.* London, Duckworth.

Crewe, R. M. 1982. Compositional variability: the key to the social signals produced by honeybee mandibular glands. Proceedings of the Ninth Congress of the International Union for Study of Social Insects, pp. 318–322.

Crewe, R. M., and H. H. W. Velthuis. 1980. False queens: a consequence of mandibular gland signals in worker honeybees. *Naturwissenschaften* 67:467–469.

Cruz-Landim, C. da. 1963. Evolution of the wax and scent glands of the Apinae. *J. N.Y. Entomol. Soc.* 71:2–13.

Cruz-Landim, C. da, and E. W. Kitajima. 1966. Ultraestrutura do aparelho venenifero de *Apis* (Hymenoptera, Apidae). *Mem. Inst. Butantan Sao Paulo* 33:701–711.

——— 1969. Ultrastructure of *Apis mellifera* hypopharyngeal gland. Proceedings of the Sixth Congress of the International Union for Study of Social Insects, pp. 121–130.

Culliney, T. W. 1983. Origin and evolutionary history of the honeybees *Apis. Bee World* 64:29–37.

Dadant, C. P., ed. 1975. *The hive and the honeybee.* Hamilton, Ill., Dadant.

Dade, H. A. 1977. *Anatomy and dissection of the honeybee.* London, International Bee Research Association.

Daly, H. V., and S. S. Balling. 1978. Identification of Africanized honeybees in the Western hemisphere by discriminant analysis. *J. Kans. Entomol. Soc.* 51:857–869.

Darchen, R. 1962. Observation directe du développement d'un rayon de cire: le role des chaînes d'abeilles. *Insectes Sociaux* 9:103–120.

——— 1968. Le travail de la cire et la construction dans la ruche. In *Traité de biologie de l'abeille,* vol. 2, ed. R. Chauvin, pp. 241–331. Paris, Masson.

DeJong, D. 1978. Insects: hymenoptera (ants, wasps, bees). In *Honey bee pests, predators, and diseases,* ed. R. A. Morse, pp. 138–157. Ithaca, Cornell Univ. Press.

Demuth, G. S. 1921. Swarm control. *Fmrs.' Bull. U.S. Dep. Agric.* 1198:1–28.

——— 1922. The cause of swarming. *Glean. Bee Cult.* 50:371–373.

——— 1931. Cause of swarming is known. *Amer. Bee J.* 71:419.

Dietz, A. 1969. Initiation of pollen consumption and pollen movement through the alimentary canal of newly emerged honey bees. *Ann. Entomol. Soc. Amer.* 62:43–46.

——— 1971. Changes with age in some mineral constituents of worker honey bees, I. Phosphorus, potassium, calcium, magnesium, sodium and iron. *J. Ga. Entomol. Soc.* 6:54–57.

Dietz, A., and W. J. Humphreys. 1971. Scanning electron microscopic studies of antennal receptors of the worker honey bee, including sensilla campaniformia. *Ann. Entomol. Soc. Amer.* 64:919–925.

Dietz, A., and E. Lambremont. 1970. Caste determination in honeybees, II. Food consumption of individual honey bee larvae, determined with 32P-labelled royal jelly. *Ann. Entomol. Soc. Amer.* 63:1342–45.

Dietz, A., R. Krell, and F. A. Eischen. 1985. Preliminary investigation on the distribution of Africanized honey bees in Argentina. *Apidologie* 16:99–108.

Dixon, S. E., and R. W. Shuel. 1963. Studies in the mode of action of royal jelly in honeybee development, III. The effect of experimental variation in diet on growth and metabolism of honeybee larvae. *Can. J. Zool.* 41:733–739.

Doedikar, G. B. 1978. Possibilities of origin and diversification of angiosperms prior to continental drift. In *Recent researches in geology,* vol. 4, ed. K. B. Powar, pp. 474–481. Delhi, Hindustan.

Doedikar, G. B., C. V. Thakar, and P. N. Shaw. 1959. Cytogenetic studies in Indian honey bees, I. Somatic chromosome complement in *Apis indica* and its bearing on evolution and phylogeny. *Proc. Ind. Acad. Sci.* 49:194–206.

Donhoff, E. 1855a. Ueber das Geruchsorgan der Biene. Repr. in *Beiträge zur Bienenkunde.* Berlin, Pfenningstorff.

——— 1855b. Ueber das Herrschen verschiedener Triebe in verschieden Lebensattern bei den Bienen. Repr. in *Beiträge zur Bienenkunde.* Berlin, Pfenningstorff.

Doolittle, R. E., M. S. Blum, and R. Boch. 1970. Cis-9-Oxo-2-decenoic acid: synthesis and evaluation as a honey bee pheromone and masking agent. *Ann. Entomol. Soc. Amer.* 63:1180–85.

Doull, K. M. 1977. Supplementary feeding of honeybees. Technical Recommendation, Beekeeping Technology and Equipment Standing Commission, Apimondia.

Drescher, W. 1969. Die Flugaktivität von Drohnen der Rasse *Apis mellifica carnica* L. und *A. mell. ligustica* L. in Abhängigkeit von Lebensalter und Witterung. *Bienenforsch.* 9:390–409.

Dreyling, L. 1903. Ueber die wachsbereitenden Organe der Honigbiene. *Zool. Anz.* 26:710–715.

Drum, N. H., and W. C. Rothenbuhler. 1984. Effect of temperature on non-stinging aggressive responses of worker honeybees to diseased and healthy bees. *J. Apic. Res.* 23:82–87.

DuPraw, E. J. 1961. A unique hatching process in the honeybee. *Trans. Amer. Micros. Soc.* 80:185–191.

——— 1967. The honeybee embryo. In *Methods in developmental biology,* ed. F. Wilt and N. Wessels, pp. 183–217. New York, Crowell.

Dutton, R. W., F. Ruttner, A. Berkeley, and M. J. D. Manley. 1981. Observations on the morphology, relationships, and ecology of *Apis mellifera* of Oman. *J. Apic. Res.* 20:201–214.

Dyer, F. C. 1985. Nocturnal orientation by the Asian honey bee, *Apis dorsata. Anim. Behav.* 33:769–774.

Dyer, F. C., and J. L. Gould. 1983. Honey bee navigation. *Amer. Sci.* 71:587–597.

Eckert, J. E. 1934. Studies in the number of ovarioles in queen honeybees in relationship to body size. *J. Econ. Entomol.* 27:627–636.

——— 1942. The pollen required by a colony of honeybees. *J. Econ. Entomol.* 35:309–311.

Eckert, J. E., and F. R. Shaw. 1960. *Beekeeping*. New York, Macmillan.

Edrich, W. 1981. Night-time sun-compass behavior of honey bees at the equator. *Physiol. Entomol.* 6:7–13.

Edrich, W., and O. von Helversen. 1976. Polarized light orientation of the honey bee: the minimum visual angle. *J. Comp. Physiol.* 109:309–314.

Erp, A. von. 1960. Mode of action of the inhibitory substance of the honeybee queen. *Insectes Sociaux* 7:207–211.

Esch, H. 1960. Üeber die Körpertemperaturen und den Wärmehaushalt von *Apis mellifica. Z. vergl. Physiol.* 43:305–355.

——— 1961. Über die Schallerzeugung beim Werbetanz der Honigbiene. *Z. vergl. Physiol.* 45:1–11.

——— 1964a. Über den Zusammenhang zwischen Temperatur, Aktionspotentiaten und Thoraxbewegungen bei der Honigbiene (*Apis mellifica*). *Z. vergl. Physiol.* 48:547–551.

——— 1964b. Beiträge zum Problem der Entfernungsweisung in den Schwänzeltänzen der Honigbiene. *Z. vergl. Physiol.* 48:534–546.

——— 1967. The sound produced by swarming honey bees. *Z. vergl. Physiol.* 56:408–411.

——— 1976. Body temperature and flight performance of honey bees in a servo-mechanically controlled wind tunnel. *J. Comp. Physiol.* 109:265–277.

Esch, H., and J. A. Bastian. 1968. Mechanical and electrical activity in the indirect flight muscles of the honey bee. *Z. vergl. Physiol.* 58:429–440.

——— 1970. How do newly recruited honey bees approach a food site? *Z. vergl. Physiol.* 68:175–181.

Esch, H., I. Esch, and W. E. Kerr. 1965. Sound: an element common to communication of stingless bees and to dances of honeybees. *Science* 149: 320–321.

Farrar, C. L. 1943. An interpretation of the problems of wintering the honeybee colony. *Glean. Bee Cult.* 71:513–518.

——— 1947. Nosema losses in package bees as related to queen supersedure and honey yields. *J. Econ. Entomol.* 40:333–338.

——— 1949. Untitled article. *Bee World* 30:51.

Fell, R. D., and R. A. Morse. 1984. Emergency queen cell production in the honey bee colony. *Insectes Sociaux* 31:221–237.

Fell, R. D., J. T. Ambrose, D. M. Burgett, D. DeJong, R. A. Morse, and T. Seeley. 1977. The seasonal cycle of swarming in honeybees. *J. Apic. Res.* 16:170–173.

Ferguson, A. W., and J. B. Free. 1979. Production of a forage-marking pheromone by the honeybee. *J. Apic. Res.* 18:128–135.

——— 1980. Queen pheromone transfer within honeybee colonies. *Physiol. Entomol.* 5:359–366.

——— 1981. Factors determining the release of Nasonov pheromone by honeybees at the hive entrance. *Physiol. Entomol.* 6:15–19.

Ferguson, A. W., J. B. Free, J. A. Pickett, and M. Winder. 1979. Techniques for studying honeybee pheromones involved in clustering, and experiments on the effect of Nasonov and queen pheromones. *Physiol. Entomol.* 4:339–344.

Filmer, R. S. 1932. Brood area and colony size as factors in activity of pollination units. *J. Econ. Entomol.* 25:336–343.

Fischer, W. 1957. Untersuchungen über die Riechschärfe der Honigbiene. *Z. vergl. Physiol.* 39:634–659.

Fletcher, D. J. C. 1975. Significance of dorsoventral abdominal vibration among honey bees (*Apis mellifera* L.). *Nature* 256:721–723.

——— 1975/1976. New perspectives in the causes of absconding in the African bee. Part I: *S. African Bee J.* 47(1975):11, 13–14. Part II: 48(1976):6–9.

——— 1977a. A preliminary analysis of rapid colony development in *Apis mellifera adansonii* L. Proceedings of the Eighth Congress of the International Union for Study of Social Insects, pp. 144–145.

——— 1977b. Evaluation of introductions of European honey bees into Southern and Eastern Africa. Proceedings of the Eighth Congress of the International Union for Study of Social Insects, pp. 146–147.

——— 1978a. The African bee, *Apis mellifera adansonii,* in Africa. *Ann. Rev. Entomol.* 23:151–171.

——— 1978b. The influence of vibratory dances by worker honeybees on the activity of virgin queens. *J. Apic. Res.* 17:3–13.

——— 1978c. Vibration of queen cells by worker honeybees and its relation to the issue of swarms with virgin queens. *J. Apic. Res.* 17:14–26.

Fletcher, D. J. C., and G. D. Tribe. 1977a. Natural emergency queen rearing by the African bee *A. m. adansonii* and its relevance for successful queen production by beekeepers, I and II. In *African bees: Taxonomy, biology and economic use,* ed. D. J. C. Fletcher, pp. 132–140, 161–168. Pretoria, Apimondia International Symposium.

——— 1977b. Swarming potential of the African bee, *Apis mellifera adansonii* L. In *African bees: taxonomy, biology and economic use,* ed. D. J. C. Fletcher, pp. 25–34. Pretoria, Apimondia International Symposium.

Forel, A. 1910. Das Sinnesleben der Insekten. Munchen: Reinhardt.

Frankland, A. W. 1976. Bee sting allergy. *Bee World* 57:145–150.

Free, J. B. 1954. The behavior of robber honeybees. *Behaviour* 7:233–240.

——— 1957a. The transmission of food between worker honeybees. *Anim. Behav.* 5:41–47.

——— 1957b. The food of adult drone honeybees (*Apis mellifera*). *Anim. Behav.* 5:7–11.
——— 1960a. The distribution of bees in a honey-bee (*Apis mellifera* L.) colony. *Proc. Roy. Entomol. Soc. London* (A) 35:141–144.
——— 1960b. The pollination of fruit trees. *Bee World* 41:141–151, 169–186.
——— 1961a. Hypopharyngeal gland development and division of labour in honey-bee (*Apis mellifera* L.) colonies. *Proc. Roy. Entomol. Soc. London* (A) 36:5–8.
——— 1961b. The stimuli releasing the stinging response of honeybees. *Anim. Behav.* 9:193–196.
——— 1963. The flower constancy of honeybees. *J. Anim. Ecol.* 32:119–131.
——— 1965. The allocation of duties among worker honeybees. *Symp. Zool. Soc. London* 14:39–59.
——— 1966. The foraging areas of honeybees in an orchard of standard apple trees. *J. Appl. Ecol.* 3:261–268.
——— 1967a. The production of drone comb by honeybee colonies. *J. Apic. Res.* 6:29–36.
——— 1967b. Factors determining the collection of pollen by honeybee foragers. *Anim. Behav.* 15:134–144.
——— 1970. The effect of flower shape and nectar guides on the behaviour of foraging honeybees. *Behaviour* 37:269–285.
——— 1977. *The social organization of the honeybees.* London, Arnold.
Free, J. B., and C. G. Butler. 1959. *Bumblebees.* London, Collins.
Free, J. B., and A. W. Ferguson. 1982. Transfer of pheromone from immature queen honeybees, *Apis mellifera. Physiol. Entomol.* 7:401–406.
Free, J. B., and P. A. Racey. 1966. The pollination of *Freesia refracta* in glasshouses. *J. Apic. Res.* 5:177–182.
Free, J. B., and J. Simpson, 1968. The alerting pheromones of the honeybee. *Z. vergl. Physiol.* 61:361–365.
Free, J. B., and I. H. Williams. 1970. Exposure of the Nasonov gland by honeybees (*Apis mellifera*) collecting water. *Behaviour* 37:286–290.
——— 1972. The role of the Nasonov gland pheromone in crop communication by honeybees (*Apis mellifera* L.). *Behaviour* 41:314–318.
——— 1975. Factors determining the rearing and rejection of drones by the honeybee colony. *Anim. Behav.* 23:650–675.
Free, J. B., and M. E. Winder. 1983. Brood recognition by honeybee (*Apis mellifera*) workers. *Anim. Behav.* 31:539–545.
Free, J. B., A. W. Ferguson, and J. A. Pickett. 1981. Evaluation of the various components of the Nasonov pheromone used by clustering honeybees. *Physiol. Entomol.* 6:263–268.
——— 1983. A synthetic pheromone lure to induce worker honeybees to consume water and artificial forage. *J. Apic. Res.* 22:224–228.
Free, J. B., J. A. Pickett, A. W. Ferguson, J. R. Simpkins, and C. Williams. 1984. Honeybee Nasonov pheromone lure. *Bee World* 65:175–181.
Free, J. B., I. H. Williams, J. A. Pickett, A. W. Ferguson, and A. P. Martin. 1982. Attractiveness of (Z)-11-eicosen-1-ol to foraging honeybees. *J. Apic. Res.* 21:151–156.

Freudenstein, H. 1960. Einfluss der Pollennahrung auf das Bauvenmögen die Wachsdrüsen und den Fettkörper der Honigbiene (*Apis mellifera* L.). *Zool. Jb. Physiol.* 69:95–124.
——— 1962. Beobachtungen über die Färbung der Waben durch die Honigbiene. *Zool. Beitr.* (Berlin) 7:311–319.
Friedmann, H. 1955. The honey-guides. *Bull. U.S. Nat. Mus.* 208:1–292.
Frisch, K. von. 1921. Über den Sitz des Geruchsinnes bei Insekten. *Zool. Jb. Physiol.* 38:1–68.
——— 1923. Ueber die "Sprache" der Bienen. *Zool. J.* 40:1–186.
——— 1934. Über den Geschmackssinn der Bienen. *Z. vergl. Physiol.* 21:1–156.
——— 1965. *Tanzsprache und Orientierung der Bienen.* Berlin, Springer-Verlag.
——— 1967a. The dance language and orientation of bees. Cambridge, Mass., Harvard Univ. Press.
——— 1967b. Honeybees: do they use direction and distance information provided by their dancers? *Science* 158:1072–76.
——— 1974. *Animal architecture.* Harcourt Brace Jovanovich. N.Y.
Frisch, K. von, and G. A. Rösch. 1926. Neue Versuche über die Bedeutung von Duftorgan und Pollenduft für die Verständigung im Bienenvolk. *Z. vergl. Physiol.* 4:1–21.
Fukuda, H. 1960. Some observations on the pollen foraging activities of the honey bee, *Apis mellifera* L. (preliminary report). *J. Fac. Sci. Hokkaido Univ.* (ser. 6) 14:381–386.
Fukuda, H., and T. Ohtani. 1977. Survival and life span of drone honeybees. *Res. Pop. Ecol.* 19:51–68.
Fukuda, H., and S. F. Sakagami. 1968. Worker brood survival in honeybees. *Res. Pop. Ecol.* 10:31–39.
Fukuda, H., and K. Sekiguchi. 1966. Seasonal change of the honeybee worker longevity in Sapporo, North Japan, with notes on some factors affecting the life span. *Jap. J. Ecol.* 16:206–212.
Fukuda, H., K. Moriya, and K. Sekiguchi. 1969. The weight of crop contents in foraging honeybee workers. *Annotnes Zool. Jap.* 42:80–90.
Furgala, B. 1962. The effect of the intensity of *Nosema* inoculum on queen supersedure in the honey bee, *Apis mellifera* Linnaeus. *J. Insect Pathol.* 4:429–432.

Garofalo, C. A. 1977. Brood viability in normal colonies of *Apis mellifera. J. Apic. Res.* 16:3–13.
Gary, N. E. 1961a. Queen honey bee attractiveness as related to mandibular gland secretion. *Science* 133:1479–80.
——— 1961b. Mandibular gland extirpation in living queen and worker honey bees (*Apis mellifera*). *Ann. Entomol. Soc. Amer.* 54:529–531.
——— 1962. Chemical mating attractants in the queen honey bee. *Science* 136:773–774.
——— 1963. Observations of mating behaviour in the honeybee. *J. Apic. Res.* 2:3–13.

——— 1966. Robbing behavior in the honey bee. *Amer. Bee J.* 106:446–448.

——— 1967. Diurnal variations in the intensity of flight activity from honeybee colonies. *J. Apic. Res.* 6:65–68.

——— 1974. Pheromones that affect the behavior and physiology of honey bees. In *Pheromones,* ed. Martin C. Birch. London, North Holland Publishing.

——— 1975. Activities and behavior of honey bees. In *The hive and the honeybee,* ed. C. P. Dadant, pp. 185–264. Hamilton, Ill., Dadant.

Gary, N. E., and J. Marston. 1971. Mating behavior of drone honey bees with queen models. *Anim. Behav.* 19:299–304.

Gary, N. E., and R. A. Morse. 1962. The events following queen cell construction in honeybee colonies. *J. Apic. Res.* 1:3–5.

Gary, N. E., P. C. Witherell, and J. M. Marston. 1972. Foraging range and distribution of honey bees used for carrot and onion pollination. *Envir. Entomol.* 1:71–78.

Gast, R. 1967. Untersuchungen über den Einfluss der Königinnensubstanz auf die Entwicklung der endokrinen Drüsen bei der Arbeiterin der Honigbiene (*Apis mellifica*). *Insectes Sociaux* 14:1–12.

Gates, B. N. 1914. The temperature of the bee colony. *Bull. U.S. Dep. Agric.* 96:1–29.

Gerig, L. 1971. Wie Drohnen auf Königinnen-attrapen reagieren. *Schweiz. Bienenztg.* 94:558–562.

——— 1972. Ein weiterer Duftstoff zur Anlockung der Drohnen von *Apis mellifica* L. *Z. Angew. Entomol.* 70:286–289.

Gerstung, F. 1891–1926. *Der Bien und Seine Zucht.* 7 edns. Berlin, Pfenningstorff.

Getz, W. M., and K. B. Smith. 1983. Genetic kin recognition: honey bees discriminate between full and half sisters. *Nature* 302:147–148.

Getz, W. M., D. Bruckner, and T. R. Parisian. 1982. Kin structure and the swarming behavior of the honey bee, *Apis mellifera. Behav. Ecol. Sociobiol.* 10:265–270.

Ghent, R. L., and N. E. Gary. 1962. A chemical alarm releaser in honey bee stings. *Psyche* 69:1–6.

Ghisalberti, E. L. 1979. Propolis: a review. *Bee World* 60:59–84.

Gillette, C. P. 1897. Weights of bees and the loads they carry. *Proc. Soc. Prom. Agric. Sci.* 14:60–63.

Goetze, G., and B. K. Bessling. 1959. Die Wirkung verschiedener Fütterung der Honigbiene auf Wachserzeugung und Bautätiglait. *Z. Bienenforsch.* 4:202–209.

Goewie, E. A. 1978. Regulation of caste differentiation in the honey bee (*Apis mellifera* L.). *Meded. Landbhoogesch. Wageningen* no. 78-15, pp. 1–75.

Gontarski, H. 1949. Ueber die Vertikalorientierung der Bienen beim Bau der Waben und bei der Anlage des Brutnestes. *Z. vergl. Physiol.* 31:652–670.

——— 1953. Zur Brutbiologie der Honegbien. *Z. Bienenforsch. Apidologia* 2:7–10.

——— 1954. Untersuchungen über die Verwertung von Pollen und Hefe zur Brutpflege der Honigbiene. *Z. Bienenforsch.* 2:161–180.

Gould, J. L. 1976. The dance-language controversy. *Q. Rev. Biol.* 51:211–244.

——— 1980. The case for magnetic sensitivity in birds and bees (such as it is). *Amer. Sci.* 68:256–268.

——— 1982. Why do honey bees have dialects? *Behav. Ecol. Soc.* 10:53–56.

Gould, J. L., J. L. Kirschvink, and K. S. Deffeyes. 1978. Bees have magnetic remanence. *Science* 201:1026–28.

Groot, A. P. de. 1951. Effect of a protein containing diet on the longevity of caged bees. *Konink. Nederl. Akad. Wetens.* 54:2–4.

——— 1953. Protein and amino acid requirements of the honeybee (*Apis mellifera* L.). *Physiol. Comp. Oecologia* 3:197–285.

Habermann, E. 1971. Chemistry, pharmacology, and toxicology of bee, wasp, and hornet venoms. In *Venomous animals and their venoms,* vol. 3 of *Venomous invertebrates,* ed. B. Cherl, W. Buckley, and E. Buckley. New York, Academic Press.

Hagerdorn, H. H., and F. E. Moeller. 1967. The rate of pollen consumption by newly emerged honeybees. *J. Apic. Res.* 6:159–162.

Halberstadt, K. 1980. Elektrophoretische Untersuch zur Sekretionstätigkeit der Hypopharynxdrüse der Honigbiene (*Apis mellifera* L.). *Insectes Sociaux* 27:61–77.

Hammann, E. 1957. Which takes the initiative in the virgin queen's flight, the queen or the workers? *Insectes Sociaux* 4:91–106.

Hanser, G., and H. Rembold. 1960. Ueber den Weiselzellenfuttersaft der Honigbiene, V. Untersuchungen über die Bildung des Futtersaftes in der Ammenbiene. *Hoppe-Seyl Z.* 319:206–212.

——— 1964. Analytische und histologische Untersuchungen der Kopf- und Thoraxdrüsen bei der Honigbiene *Apis mellifica. Z. Naturforsch.* 19:938–943.

Harbo, J. R. 1979. The rate of depletion of spermatozoa in the queen honeybee spermatheca. *J. Apic. Res.* 18:204–207.

Harbo, J. R., A. B. Bolten, T. E. Rinderer, and A. M. Collins. 1981. Development periods for eggs of Africanized and European honeybees. *J. Apic. Res.* 20:156–159.

Harrison, J. M. 1986. Caste-specific changes in honeybee flight capacity. *Physiol. Zool.* 59:175–187.

Hassanein, M. H., and M. A. El-Banby. 1960. Studies on the brood rearing activity of certain races of the honeybee *Apis mellifera* L. *Bull. Soc. Entomol. Egypte* 44:13–22.

Haydak, M. H. 1934. Changes in total nitrogen content during the life of the imago of the worker honeybee. *J. Apic. Res.* 49:21–28.

——— 1935. Brood rearing by honeybees confined to pure carbohydrate diet. *J. Econ. Entomol.* 28:657–660.

——— 1937a. Further contribution to the study of pollen substitutes. *J. Econ. Entomol.* 30:637–642.

——— 1937b. The influence of a pure carbohydrate diet on newly emerged honey bees. *Ann. Entomol. Soc. Amer.* 30:258–262.

——— 1943. Larval food and development of castes in the honeybee. *J. Econ. Entomol.* 36:778–792.

——— 1951. How long does a bee live after losing its sting? *Glean. Bee Cult.* 79:85–86.

——— 1954. Propolis. Report of the Iowa State Apiarist, pp. 74–87.

——— 1957. The food of the drone larvae. *Ann. Entomol. Soc. Amer.* 50:73–75.

——— 1958. Wintering of bees in Minnesota. *J. Econ. Entomol.* 51:332–334.

——— 1970. Honey bee nutrition. *Ann. Rev. Entomol.* 15:143–156.

Haydak, M. H., and A. Dietz. 1965. Influence of the diet on the development and brood rearing of honeybees. *Proc. Int. Beekeeping Congr., Bucharest,* 20:158–162.

——— 1972. Cholesterol, pantothenic acid, pyridoxine and thiamine requirements of honeybees for brood rearing. *J. Apic. Res.* 11:105–109.

Haydak, M. H., and A. E. Vivino. 1950. The changes in the thiamine, riboflavin, niacin, and pantothenic acid content in the food of female honeybees during growth with a note on the vitamin K activity of royal jelly and beebread. *Ann. Entomol. Soc. Amer.* 43:361–367.

Hazelhoff, E. H. 1941. De Luchtverversching van een Bijenkast gedurende den zomer. *Maandschr. Bijent.* 44:1–16.

Heberle, J. A. 1914. Notes from Germany: how many trips to the field does a bee make in a day? How long does it take to fetch one load? How long does a bee remain in the hive between trips? *Glean. Bee Cult.* 42:904–905.

Heinrich, B. 1979a. *Bumblebee economics.* Cambridge, Mass., Harvard Univ. Press.

——— 1979b. Keeping a cool head: honeybee thermoregulation. *Science* 205:1269–71.

——— 1979c. Thermoregulation of African and European honeybees during foraging, attack, and hive exits and returns. *J. Exp. Biol.* 80:217–229.

——— 1980a. Mechanisms of body-temperature regulation in honeybees, *Apis mellifera,* I. Regulation of head temperature. *J. Exp. Biol.* 85:61–72.

——— 1980b. Mechanisms of body-temperature regulation in honeybees. *Apis mellifera,* II. Regulation of thoracic temperature at high air temperatures. *J. Exp. Biol.* 85:73–87.

——— 1981a. Energetics of honeybee swarm thermoregulation. *Science* 212:565–566.

——— 1981b. The regulation of temperature in the honeybee swarm. *Sci. Amer.* 244:147–160.

——— 1981c. The mechanisms and energetics of honeybee swarm temperature regulation. *J. Exp. Biol.* 91:25–55.

——— 1985. The social physiology of temperature regulation in honeybees. *Fortsch. Zool.* 31:393–406.

Hellmich, R. L., J. M. Kulincevic, and W. C. Rothenbuhler. 1985. Selection for high and low pollen-hoarding honey bees. *J. Heredity* 76:155–158.

Hemmling, C., N. Koeniger, and F. Ruttner. 1979. Quantitative Bestimmung der 9-Oxodecensäure im Lebenszyklus der Kapbiene (*Apis mellifera capensis* Escholtz). *Apidologie* 10:227–240.

Hemstedt, H. 1969. Zum Feinbau der Koshewnikowschen Drüse bei der Honigbiene *Apis mellifica* (Insecta, Hymenoptera). *Z. Morph. Tiere* 66:51–72.

Heran, H. 1952. Untersuchungen über den Temperatursinn der Honigbiene (*Apis mellifica*) unter besonderer Berücksichtigung der Wahrnehmung strahlender Wärme. *Z. vergl. Physiol.* 34:179–206.

——— 1959. Wahrnehmung und Regelung der Flugeigengeschwindigkeit bei *Apis mellifica* L. *Z. vergl. Physiol.* 42:103–163.

Herbert, E. W., Jr., J. T. Vanderslice, and D. J. Higgs. 1985. Vitamin C enhancement of brood rearing by caged honeybees fed a chemically defined diet. *Arch. Insect Biochem. Physiol.* 2:29–37.

Heselhaus, F. 1922. Die Hautdrusen der Apiden und verwandter Formen. *Zool. Jb.*, Abt. 2, 43:369–464.

Hess, W. R. 1926. Die Temperaturregulierung im Bienenvolk. *Z. vergl. Physiol.* 4:465–487.

Himmer, A. 1927. Der soziale Wärmehaushalt der Honigbiene, II, Die Wärme der Bienenbrut. *Erlanger Jb. Bienenk.* 5:1–32.

Hirschfelder, H. 1951. Quantitative Untersuchungen zum Polleneintragen der Bienenvölker. *Z. Bienenforsch.* 1:67–77.

Hodges, D. 1952. *The pollen loads of the honeybee.* London, Bee Research Association.

——— 1967. The function of the single hair on the floor of the honeybee's corbicula. *Bee World* 48:58–61.

Hopkins, C. Y., A. W. Jevans, and R. Boch. 1969. Occurrence of octadeca-trans-2, cis-12-trienoic acid in pollen attractive to the honey bee. *Can. J. Biochem.* 47:433–436.

Howell, D. E., and R. L. Usinger. 1933. Observations on the flight and length of life of drone bees. *Ann. Entomol. Soc. Amer.* 26:239–246.

Huber, F. 1792. *New observations on bees,* vol. 1. Transl. (1926), Hamilton, Ill., Dadant.

Imboden, H., and M. Lüscher. 1975. Allatektomie bei adulten Bienen-Arbeiterinnen (*Apis mellifica*). *Rev. Swisse Zool.* 82:694–698.

Istomina-Tsvetkova, K. P. 1953. Reciprocal feeding of bees. *Pchelovodstvo* 30:25–29 (in Russian).

Jack, R. W. 1916. Parthenogenesis amongst the workers of the Cape honeybee: Mr. G. W. Onion's experiments. *Trans. Roy. Entomol. Soc. London* 64: 396–403.

Jacobs, W. 1924. Das Duftorgan von *Apis mellifica* und ähnliche Hautdrüsenorgane sozialer und solitürer Apiden. *Zeitschr. Morph. Ökol. Tiere* 3:1–80.

Jander, R. 1976. Grooming and pollen manipulation in bees (Apoidea): the nature and evolution of movements involving the foreleg. *Physiol. Entomol.* 1:179–194.

Jay, S. C. 1962a. Prepupal and pupal ecdyses of the honeybee. *J. Apic. Res.* 1:14–18.

——— 1962b. Colour changes in honeybee pupae. *Bee World* 43:119–122.

——— 1963a. The development of honeybees in their cells. *J. Apic. Res.* 2:117–134.

——— 1963b. The longitudinal orientation of the larval honeybees (*Apis mellifera*) in their cells. *Can. J. Zool.* 41:717–723.

——— 1964a. The cocoon of the honeybee, *Apis mellifera* L. *Can. Entomol.* 96:784–792.

——— 1964b. Starvation studies of larval honey bees. *Can. J. Zool.* 42:455–462.

——— 1970. The effect of various combinations of immature queen and worker bees on the ovary development of worker honeybees in colonies with and without queens. *Can. J. Zool.* 48:169–173.

——— 1972. Ovary development of worker honeybees when separated from worker brood by various methods. *Can. J. Zool.* 50:661–664.

——— 1986. Spatial management of honey bees on crops. *Ann. Rev. Entomol.* 31:49–65.

Jaycox, E. R. 1956. Factors affecting the attainment of sexual maturity by the drone honeybee (*Apis mellifera* L.) Ph.D. diss., Univ. of Calif., Davis.

——— 1970a. Honey bee queen pheromones and worker foraging behavior. *Ann. Entomol. Soc. Amer.* 63:222–228.

——— 1970b. Honey bee foraging behavior: responses to queens, larvae, and extracts of larvae. *Ann. Entomol. Soc. Amer.* 63:1689–94.

——— 1976. Behavioral changes in worker honey bees (*Apis mellifera* L.) after injection with synthetic juvenile hormone (Hymenoptera: Apidae). *J. Kans. Entomol. Soc.* 49:165–170.

Jaycox, E. R., and S. G. Parise. 1980. Homesite selection by Italian honey bee swarms *Apis mellifera ligustica* (Hymenoptera: Apidae). *J. Kans. Entomol. Soc.* 53:171–178.

——— 1981. Homesite selection by swarms of black-bodied honeybees, *Apis mellifera caucasica* and *A. m. carnica* (Hymenoptera: Apidae). *J. Kans. Entomol. Soc.* 54:697–703.

Jaycox, E. R., W. Skowronek, and G. Gwynn. 1974. Behavioral changes in worker honey bees (*Apis mellifera*) induced by injections of a juvenile hormone mimic. *Ann. Entomol. Soc. Amer.* 67:529–534.

Jeffree, E. P. 1951. The swarming period in Wiltshire. *Wiltshire Beekprs. Gaz.* 74:2–3.

——— 1955. Observations on the decline and growth of honey bee colonies. *J. Econ. Entomol.* 48:723–726.

——— 1956. Winter brood and pollen in honey bee colonies. *Insectes Sociaux* 3:417–422.

Johnson, D. L. 1967. Honeybees: do they use the direction information contained in their dance maneuver? *Science* 155:844–847.

Jongbloed, J., and C. A. G. Wiersma. 1934. Der Stoffwechsel der Honigbiene während des Fliegens. *Z. vergl. Physiol.* 21:519–533.

Jung-Hoffmann, I. 1966. Die Determination von Königin und Arbeiterin der Honigbiene. *Z. Bienenforsch.* 8:296–322.

——— 1967. Brutpflege und Königinentstehung. *Imkerfreund* 22:111–113.

Juska, A. 1978. Temporal decline in attractiveness of honeybee queen tracks. *Nature* 276:261.

Juska, A., T. D. Seeley, and H. H. W. Velthuis. 1981. How honeybee queen attendants become ordinary workers. *J. Insect Physiol.* 27:515–519.

Kaiser, W. 1972. A preliminary report of the analysis of the optomotor system of the honey bee: single unit recordings during stimulation with spectral light. In *Information processing in the visual systems of arthropods,* ed. R. Wehner, pp. 166–170. New York, Springer-Verlag.

Kaissling, K. E., and M. Renner. 1968. Antennale Rezeptoren fur Queen Substance and Sterzelduft bei der Honigbiene. *Z. vergl. Physiol.* 59:357–361.

Kalmus, H., and C. R. Ribbands. 1952. The origin of the odours by which honeybees distinguish their companions. *Proc. Roy. Soc. London* (B) 140:50–59.

Kelner-Pillault, S. 1969. Abeilles fossiles ancetres des Apides sociaux. Proceedings of the Sixth Congress of the International Union for Study of Social Insects, pp. 85–93.

Kerr, W. E. 1967. Multiple alleles and genetic load in bees. *J. Apic. Res.* 6:61–63.

Kerr, W. E., and N. J. Hebling. 1964. Influence of the weight of worker bees on division of labor. *Evolution* 18:267–270.

Kerr, W. E., and E. de Lello. 1962. Sting glands in stingless bees—a vestigial character. *J. N.Y. Entomol. Soc.* 70:190–214.

Kerr, W. E., S. L. D. Rio, and M. D. Barrionuevo. 1982. The southern limits of the distribution of the Africanized honey bee in South America. *Am. Bee J.* 122:196, 198.

Kerr, W. E., M. Blum, J. Pisani, and A. C. Stort. 1974. Correlation between amounts of 2-heptanone and iso-amyl acetate in honeybees and their aggressive behaviour. *J. Apic. Res.* 13:173–176.

Kerr, W. E., L. S. Goncalves, L. F. Blotta, and H. B. Maciel. 1972. Biologia comparada entre as abelhas italianas (*Apis mellifera ligustica*), africana (*A. m. adansonii*) e suas hibridas. Proceedings of the First Brazilian Congress of Apiculture, pp. 151–185.

Kerr, W. E., R. Zucchi, J. T. Nakadaira, and J. E. Butolo. 1962. Reproduction in the social bees (Hymenoptera: Apidae). *J. N.Y. Entomol. Soc.* 70:265–276.

Keularts, J. L., and H. F. Linskens. 1968. Influence of fatty acids on *Petunia* pollen grains. *Acta Botan. Neerl.* 17:267–272.

Kiechle, H. 1961. Die soziale Regulation der Wassersammeltätigkeit im Bienenstaat und deren physiologische Grundlage. *Z. vergl. Physiol.* 45:154–192.

Kimsey, L. S. 1984. A re-evaluation of the phylogenetic relationships in the Apidae (H). *Syst. Entomol.* 9:435–441.

King, G. E. 1933. The larger glands in the worker honey-bee: a correlation of activity with age and with physiological functioning. Ph.D. diss., Univ. of Ill., Urbana.

Kinoshita, G., and R. W. Shuel. 1975. Mode of action of royal jelly in honeybee development, X. Some aspects of lipid nutrition. *Can. J. Zool.* 53:311–319.

Klungness, L. M., and Y. S. Peng. 1984. Scanning electron microscope observations of pollen food bolus in the alimentary canal of honeybees (*Apis mellifera* L.). *Can. J. Zool.* 62:1316–19.

Knaffl, H. 1953. Über die Flugweite und Entfernungsmeldung der Bienen. *Z. Bienenforsch.* 2:131–140.

Knox, D. A., H. Shimanuki, and E. W. Herbert. 1971. Diet and the longevity of adult honey bees. *J. Econ. Entomol.* 64:1415–16.

Koeninger, G. 1984. Funktionsmorphologische Befunde bei der Kopulation der Honigbiene (*Apis mellifera* L.). *Apidologie* 15:189–204.

Koeninger, G., N. Koeniger, and M. Fabritius. 1979. Some detailed observations of mating in the honeybee. *Bee World* 60:53–57.

Koeniger, N. 1969. Experiments concerning the ability of the queen (*Apis mellifica* L.) to distinguish between drone and worker cells. Summary of the Twenty-Second International Beekeeping Congress, p. 138.

——— 1970. Factors determining the laying of drone and worker eggs by the queen honeybee. *Bee World* 51:166–169.

Koeniger, N., and H. J. Veith. 1983. Glyceryl-1,2-dioleate-3-palmitate as a brood pheromone of the honey bee (*Apis mellifera* L.). *Experientia* 39: 1051–57.

Kolmes, S. A. 1985a. An ergonomic study of *Apis mellifera* (Hymenoptera:Apidae). *J. Kans. Entomol. Soc.* 58:413–421.

——— 1985b. An information-theory analysis of task specialization among worker honey bees performing hive duties. *Anim. Behav.* 33:181–187.

——— 1985c. A quantitative study of the division of labour among worker honey bees. *Z. Tierpsychol.* 68:287–302.

Koptev, V. S. 1957. Laying workers and swarming. *Pchelovodstvo* 34:31–32 (in Russian).

Korst, P. J. A. M., and H. H. W. Velthuis. 1982. The nature of trophallaxis in honeybees. *Insectes Sociaux* 29:209–221.

Kosmin, N. P., W. W. Alpatov, and M. S. Resnitschenko. 1932. Zur Kenntnis des Gaswechsels und Energieverbrauchs der Biene in Beziehung zu deren Aktivität. *Z. vergl. Physiol.* 17:408–422.

Kramer, E. 1976. The orientation of walking honeybees in odour fields with smell concentration gradients. *Physiol. Entomol.* 1:27–37.

Kratky, E. 1931. Morphologie und Physiologie der Drusen in Kopf und Thorax der Honigbiene. *Z. wiss. Zool.* 139:120–200.

Kronenberg, F., and H. C. Heller. 1982. Colonial thermoregulation in honey bees (*Apis mellifera*). *J. Comp. Physiol.* 148:65–76.

Kroon, G. H., J. P. van Praagh, and H. H. W. Velthuis. 1974. Osmotic shock as a prerequisite to pollen digestion in the alimentary tract of the worker honeybee. *J. Apic. Res.* 13:177–181.

Kropacova, S., and H. Haslbachova. 1970. The development of ovaries in worker honeybees in queenright colonies examined before and after swarming. *J. Apic. Res.* 9:65–70.

——— 1971. The influence of queenlessness and of unsealed brood on the development of ovaries in worker honeybees. *J. Apic. Res.* 10:57–61.

Kulincevic, J. M., and W. C. Rothenbuhler. 1982. Selection for length of life in the honeybee (*Apis mellifera*). *Apidologie* 13:347–352.

Kuterbach, D. A., B. Walcott, R. J. Reeder, and R. B. Frankel. 1982. Iron-containing cells in the honey bee (*Apis mellifera*). *Science* 218:695–697.

Kuwabara, M. 1947. Ueber die Regulation im weisellosen Volke der Honigbiene, besonders die Bestimmung des neuen Weisels. *J. Fac. Sci. Hokkaido Univ.* (ser. 6) 9:359–381.

Kuwabara, M., and K. Takeda. 1956. On the hygroreceptor of the honeybee *Apis mellifera. Physiol. Ecol.* 7:1–6.

Lacher, V. 1964. Elektrophysiologische Untersuchungen an einzelnen Rezeptoren für Geruch, Luftfeuchtigkeit und Temperatur auf den Antennen der Arbeitsbienen und der Drohne. *Z. vergl. Physiol.* 48:587–623.

Lacher, V., and D. Schneider. 1963. Elektrophysiologischer Nachweis der Riechfunktion von Porenplatten (Sensilla placodea) auf den Antennen der Drohne und der Arbeitsbiene (*Apis mellifica* L.). *Z. vergl. Physiol.* 47:274–278.

Laidlaw, H. H. 1944. Artificial insemination of the queen bee (*Apis mellifera* L.): morphological basis and results. *J. Morphol.* 74:429–465.

Laidlaw, H. H., and R. E. Page. 1984. Polyandry in honey bees (*Apis mellifera* L.): sperm utilization and intracolony genetic relationships. *Genetics* 108:985–997.

Laidlaw, H. H., F. P. Gomes, and W. E. Kerr. 1956. Estimation of the number of lethal alleles in a panmitic population of *Apis mellifera* L. *Genetics* 41:179–188.

Lavie, P. 1960. Les substances antibactériennes dans la colonie d'abeilles (*Apis mellifica* L.). *Ann. Abeille* 3:103–183, 201–305.

Lecomte, J. 1957. Über die Bildung von "Strassen" durch Sammelbienen, deren stock um 180° gedrecht worde. *Z. Bienenforsch.* 3:128–133.

Lee, P. C. 1985. Reproduction and growth of temperate-evolved honey bee colonies (*Apis mellifera* L.). M.Sc. thesis, Simon Fraser Univ., Burnaby, B.C.

Lee, P. C., and M. L. Winston, 1985a. The influence of swarm population on brood production and emergent worker weight in newly founded honey bee colonies (*Apis mellifera*). *Insectes Sociaux* 32:96–103.

——— 1985b. The effect of swarm size and date of issue on comb construction in newly founded colonies of honey bees (*Apis mellifera* L.) *Can. J. Zool.* 63:524–527.

Lensky, Y., and H. Siefert. 1980. The effect of volume, ventilation, and overheating of bee colonies on the construction of swarming queen cups and cells. *Comp. Biochem. Physiol.* 67a:97–101.

Lensky, Y., and H. Skolnik. 1980. Immunochemical and electrophoretic identification of the vitellogenin proteins of the queen bee (*Apis mellifera*). *Comp. Biochem. Physiol.* 66:185–193.

Lensky, Y., and Y. Slabezki. 1981. The inhibiting effect of the queen bee (*Apis mellifera* L.) foot-print pheromone on the construction of swarming queen cups. *J. Insect Physiol.* 27:313–323.

Lensky, Y., J. C. Baehr, and P. Porcheron. 1978. Dosages radio-immunologiques des ecdysones et des hormones juvéniles au cours du développement post-embryonnaire chez les ouvrières et les reines d'abeille (*Apis mellifera* L. var. ligustica). *C. R. Acad. Sci.* 287:821–824.

Lensky, Y., P. Cassier, M. Notkin, C. Delorme-Joulie, and M. Levinsohn.

1985. Pheromonal activity and fine structure of the mandibular glands of honeybee drones (*Apis mellifera* L.) (Insecta, Hymenoptera, Apidae). *J. Insect Physiol.* 31:265–276.

Levin, M. D. 1966. Orientation of honeybees in alfalfa with respect to landmarks. *J. Apic. Res.* 5:121–125.

Levin, M. D., and S. Glowska-Konopacka. 1963. Responses of foraging honeybees in alfalfa to increasing competition from other colonies. *J. Apic. Res.* 2:33–42.

Lindauer, M. 1949. Ueber die Einwirkung von Duft- und Geschmacksstoffen sowie anderer Faktoren auf die Tänze der Bienen. *Z. vergl. Physiol.* 31: 348–412.

——— 1951. Bee dances in the clustered swarm. *Naturwissenschaften* 22:509–513.

——— 1952. Ein Beitrag zur Frage der Arbeitsteilung im Bienenstaat. *Z. vergl. Physiol.* 34:299–345. (Transl. *Bee World* 34:63–73, 85–90.)

——— 1953. Division of labour in the honeybee colony. *Bee World* 34:63–73, 85–91.

——— 1954. Temperaturregulierung und Wasserhaushalt im Bienenstaat. *Z. vergl. Physiol.* 36:391–432.

——— 1955. Schwarmbienen auf Wohnungssuche. *Z. vergl. Physiol.* 37:263–324.

——— 1961. *Communication among social bees.* Cambridge, Mass., Harvard Univ. Press.

Lindauer, M., and H. Martin. 1963. Über die Orientierung der Biene im Duftfeld. *Naturwissenschaften* 50:509–514.

Lindauer, M., and B. Schricker. 1963. Über die Funktion der Ocellen bei den Dämmerungsflügen der Honigbiene. *Biol. Zbl.* 82:721–725.

Lineburg, B. 1923a. What do bees do with brood cappings? *Amer. Bee J.* 63:235–236.

——— 1923b. Conservation of wax by the bees. *Amer. Bee J.* 63:615–616.

——— 1924. The feeding of honeybee larvae. *Bull. U.S. Dep. Agric.* 1222:25–37.

Lingens, F., and H. Rembold. 1959. Ueber den Weiselzellenfuttersaft der Honigbiene, III. Vitamingehalt von Koniginnen- und Arbeitsteilung im Bienenstaat. *Z. vergl. Physiol.* 34:299–345.

Loper, G. M. 1985. Influence of age on the fluctuation of iron in the oenocytes of honey bee (*Apis mellifera*) drones. *Apidologie* 16:181–184.

Louveaux, J. 1958. Recherches sur la récolte du pollen par les abeilles (*Apis mellifica* L.). *Ann. Abeille* 1:113–188, 197–221.

——— 1966. Les modalités de l'adaptation des Abeilles (*Apis mellifica* L.) au milieu naturel. *Ann. Abeille* 9:323–350.

Lukoschus, F. 1955. Uber Hautsinnesorgane der Bienenlarvae (*Apis mellifera*). *Z. Bienenforsch.* 3:85–87.

——— 1956. Untersuchungen zur Entwicklung der Kasten-Merkmale bei der Honigbiene (*Apis mellifica* L.) *Zeitschr. Morphol. Ökol. Tiere* 45:157–197.

Lundie, A. E. 1925. The flight activities of the honeybee. *Bull. U.S. Dept. Agric.* 1328:1–38.

Lüscher, M., and I. Walker. 1963. Zur Frage der Wirkungsweise der Königinnenpheromone bei der Honigbiene. *Rev. Suisse Zool.* 70:304–311.

McIndoo, N. E. 1914. The scent-producing organ of the honeybee. *Proc. Acad. Nat. Sci.* (Philadelphia) 66:542–555.

Mackensen, O. 1943. The occurrence of parthenogenetic females in some strains of honeybees. *J. Econ. Entomol.* 36:465–467.

——— 1951. Viability and sex determination in the honey bee (*Apis mellifera* L.). *Genetics* 36:500–509.

——— 1955. Further studies on a lethal series in the honey bee. *J. Heredity* 46:72–74.

Mackensen, O., and W. P. Nye. 1966. Selecting and breeding honey bees for collecting alfalfa pollen. *J. Apic. Res.* 5:79–86.

——— 1969. Selective breeding of honey bees for alfalfa pollen collection: sixth generation and outcrosses. *J. Apic. Res.* 8:9–12.

Mackensen, O., and W. C. Roberts. 1948. A manual for the artificial insemination of queen bees. *U.S. Dep. Agric. Bur. Entomol. Plant Q.*, ET–250.

Malyshev, S. I. 1968. *Genesis of the Hymenoptera.* London, Methuen.

Manning, F. J. 1952. Recent and fossil honeybees: some aspects of their cytology, phylogeny and evolution. *Proc. Linn. Soc. Lond.* 163:3–8.

——— 1960. A new fossil bee from Baltic amber. *Proc. XI Int. Congr. Entomol.* (Vienna) 1:306–308.

Marchand, C. 1967. Préparons le piegeage des essaims. *Abeille Fr.* 46:59–61.

Marshall, J. 1935a. On the sensitivity of the chemoreceptors of the antenna and fore-tarsus of the honeybee. *J. Exp. Biol.* 12:17–26.

——— 1935b. The location of olfactory receptors in insects: a review of experimental evidence. *Trans. Roy. Entomol. Soc. London* 83:49–72.

Martin, H. 1964. Zur Nahorientierung der Biene im Duftfeld zugleich ein Nachweis für die Osmotropotaxis bei Insekten. *Z. vergl. Physiol.* 48:481–553.

Martin, H., and M. Lindauer. 1966. Sinnesphysiologische Leistungen beim Wabenbau der Honigbiene. *Z. vergl. Physiol.* 53:372–404.

Martin, P. 1963. Die Steuerung der Volksteilung beim Schwärmen der Bienen: zugleich ein Beitrag zum Problem der Wanderschwärme. *Insectes Sociaux* 10:13–42.

Maschwitz, U. W. 1964a. Alarm substances and alarm behaviour in social Hymenoptera. *Nature* 204:324–327.

——— 1964b. Gefahrenalarmstoffe und Gefahrenalarmierung bei sozialen Hymenopteren. *Z. vergl. Physiol.* 47:596–655.

Masson, C., and G. Arnold. 1984. Ontogeny, maturation and plasticity of the olfactory system in the worker bee. *J. Insect Physiol.* 30:7–14.

Matsuka, M., N. Watabe, and K. Takeuchi. 1973. Analysis of the food of larval drone honeybees. *J. Apic. Res.* 12:3–7.

Maurizio, A. 1950. The influence of pollen feeding and brood rearing on the length of life and physiological condition of the honeybee. *Bee World* 31:9–12.

——— 1953. Weitere Untersuchungen an Pollenhöschen, Beitrag zur Erfassung der Pollentractverhaltnisse in verschiedenen Gegenden der Schweiz, Schweizerische Bienen-zeitung. *Beihefte* 2:485–556.

——— 1954. Pollen: its composition, collection, utilization, and identification. *Bee World* 35:49–50.

——— 1959. Factors influencing the lifespan of bees. Proceedings of the Ciba Foundation Symposium, pp. 231–243.

——— 1960. Bestimmung der letalen Dosis einiger Flourverbindungen für Bienen: zugleich ein Beitrag für Methodik der Giftwertbestimmung in Bienenversuchen. *IV Intern. Congr. Plant Protect. 1957,* 2:1709–1713.

——— 1975. How bees make honey. In *Honey, a comprehensive survey,* ed. E. Crane, pp. 77–105. London, Heinemann.

Mautz, D. 1971. Der Kommunikationseffekt der Schwänzeltänze bei *Apis mellifica carnica* (Pollm.). *Z. vergl. Physiol.* 72:197–220.

Melampy, R. M., and Willis, E. R. 1939. Respiratory metabolism during larval and pupal development of the female honeybee. *Physiol. Zool.* 12: 302–311.

Menzel, R. 1973. Spectral responses of moving detecting and "sustained" fibres in the optic lobe of the bee. *J. Comp. Physiol.* 82:135–150.

Menzel, R., J. Erber, and T. Masuhr. 1973. Learning and memory in the honey bee. In *Experimental analysis of insect behavior,* ed. L. B. Browne, pp. 195–217. New York, Springer-Verlag.

Merrill, J. H. 1924. Observations on brood-rearing. *Amer. Bee J.* 64:337–338.

Meyer, W. 1954. Die "Kittharzbienen" und ihre Tätigkeiten. *Z. Bienenforsch.* 5:185–200.

——— 1956a. "Propolis bees" and their activities. *Bee World* 37:25–36.

——— 1956b. Arbeitsteilung im Bienenschwarm. *Insectes Sociaux* 3:303–324.

Meyer, W., and W. Ulrich. 1952. Zur Analyse der Bauinstinkte unserer Honigbiene: untersuchungen über die Kleinbauarbeiten. *Naturwissenschaften* 39:264.

Michelsen, A., W. H. Kirchner, and M. Lindauer. 1986. Sound and vibrational signals in the dance language of the honeybee, *Apis mellifera. Behav. Ecol. Sociobiol.* 18:207–212.

Michener, C. D. 1963a. The establishment and spread of European honey bees in Australian New Guinea. *Bee World* 44:81.

——— 1963b. A further note on *Apis mellifera* in New Guinea. *Bee World* 44:114.

——— 1969. Comparative social behavior of bees. *Ann. Rev. Entomol.* 14:299–342.

——— 1974. *The social behavior of the bees: a comparative study.* Cambridge, Mass., Harvard Univ. Press.

Michener, C. D., and R. W. Brooks. 1984. Comparative study of the glossae of bees. *Contrib. Amer. Entomol. Inst.* 22:1–73.

Michener, C. D., and A. Fraser. 1978. A comparative anatomical study of mandibular structures in bees. *Univ. Kans. Sci. Bull.* 51:463–482.

Michener, C. D., and L. Greenberg. 1980. Ctenoplectridae and the origin of long-tongued bees. *Zool. J. Linn. Soc.* 69:183–203.

Michener, C. D., M. L. Winston, and R. Jander. 1978. Pollen manipulation and related activities and structures in the Apidae. *Univ. Kans. Sci. Bull.* 51:575–601.

Miller, C. C. 1902. Untitled article. *Glean Bee Cult.* 30:136.

Milojevic, B. D. 1940. A new interpretation of the social life of the honeybee. *Bee World* 21:39–41.

Milum, V. G. 1955. Honey bee communication. *Amer. Bee J.* 95:97–104.

Mindt, B. 1962. Untersuchungen über das Leben der Drohnen, insbesondere Ernahrung und Geschlechtsreife. *Z. Bienenforsch.* 6:9–33.

Minnich, D. E. 1932. The contact chemoreceptors of the honeybee. *J. Exp. Zool.* 61:375–393.

Mitchell, C. 1970. Weights of workers and drones. *Amer. Bee J.* 110:468–469.

Moeller, F. E. 1972. Honey bee collection of corn pollen reduced by feeding pollen in the hive. *Amer. Bee J.* 112:210–212.

Moore, A. J., M. Breed, and M. J. Moor. 1987. Characterization of guard behavior in honey bees, *Apis mellifera*. *Anim. Behav.* (in press).

Moore, D., J. Penikas, and M. A. Rankin. 1981. Regional specialization for an optomotor response in the honeybee compound eye. *Physiol. Entomol.* 6:61–69.

Morales, G. 1986. Effects of cavity size on demography of unmanaged colonies of honey bees (*Apis mellifera* L.). M.Sc. thesis, Univ. of Guelph, Ontario.

Moritz, R. F. A. 1983. Homogeneous mixing of honeybee semen by centrifugation. *J. Apic. Res.* 22:249–255.

——— 1985. The effects of multiple mating on the worker-queen conflict in *Apis mellifera* L. *Behav. Ecol. Sociobiol.* 16:375–377.

Moritz, R. F. A., and D. Kauhausen. 1984. Hybridization between *Apis mellifera capensis* and adjacent races of *Apis mellifera*. *Apidologie* 15:211–222.

Morland, D. 1930. On the causes of swarming in the honeybee: an examination of the brood food theory. *Ann. Appl. Biol.* 17:137–147.

Morse, R. A. 1963. Swarm orientation in honeybees. *Science* 141:357–358.

Morse, R. A., ed. 1978. *Honey bee pests, predators, and diseases.* Ithaca, Cornell Univ. Press.

Morse, R. A., and R. Boch. 1971. Pheromone concert in swarming honey bees. *Ann. Entomol. Soc. Amer.* 64:1414–17.

Morse, R. A., and T. D. Seeley. 1978. Bait hives. *Glean. Bee Cult.* 106:218–220, 242.

Morse, R. A., G. E. Strang, and J. Nowakowski. 1967. Fall death rates of drone honey bees. *J. Econ. Entomol.* 60:1198–1202.

Morton, K. 1950. The food of worker bees of different ages. *Bee World* 32:78–79.

Moskovljevic, V. 1940. Untitled report. *Bee World* 21:39–41.

Moskovljevic-Filipovic, V. C. 1956. The influence of normal and modified social structures on the development of the pharyngeal glands and the work of the honeybee (*Apis mellifera* L.). *Monogr. Srpska. Akad. Nauka* 262:101 (in Russian).

Müller, E. 1950. Ueber Drohnensammelplätze. *Bienenvater, Wien* 75:264–265.

Murray, J. A. 1964. A case of multiple bee stings. *Cent. Afr. J. Med.* 10:249–251.

Murray, L., and E. P. Jeffree. 1955. Swarming in Scotland. *Scottish Beekpr.* 31:96–98.

Myser, W. C. 1952. Ingestion of eggs by honey bee workers. *Amer. Bee J.* 92:67.

Nedel, J. O. 1960. Morphologie und Physiologie der Mandebeldrüse einiger Bienen-Arten (Apidae). *Z. Morphol. Ökol. Tiere* 49:139–183.

Neese, V. 1965. Zur Funktion der Augenborsten bei der Honigbiene. *Z. vergl. Physiol.* 49:543–585.

Nelson, D. L., and N. E. Gary. 1983. Honey productivity of honeybee colonies in relation to body weight, attractiveness and fecundity of the queen. *J. Apic. Res.* 22:209–213.

Neukirch, A. 1982. Dependence of the life span of the honeybee (*Apis mellifera*) upon flight performance and energy consumption. *J. Comp. Physiol.* 146:35–40.

Nightingale, J. 1976. Traditional beekeeping among Kenya tribes, and methods proposed for improvement and modernisation. In *Apiculture in tropical climates*, ed. E. Crane, pp. 15–22. London, International Bee Research Association.

Nixon, H. L., and C. R. Ribbands. 1952. Food transmission within the honeybee community. *Proc. Roy. Soc. London* (B) 140:43–50.

Nogueira, R. H. F., and L. S. Goncalves. 1982. Study of gland size and type in *Apis mellifera* workers emerged from drone cells. *Rev. Brasil. Genética* 5:51–59.

Nolan, W. J. 1924. The division of labor in the honeybee. *N.C. Beekpr.* (Oct.):10–15.

——— 1925. The brood-rearing cycle of the honeybee. *Bull. U.S. Dep. Agric.* 1349:1–56.

——— 1928. Seasonal brood rearing activity of the Cyprian honeybee. *J. Econ. Entomol.* 21:392–403.

Nowakowski, J. 1969. Relation between the volume of a comb cell and the number of generations and weight of workers reared in it. *Zesz. nauk. wyzsz Szk. roln. Wrocl.* 16:65–73 (in Polish, with an English summary).

Nowogrodzki, R. 1984. Division of labor in the honeybee colony: a review. *Bee World* 65:109–116.

Nuñez, J. A. 1979. Time spent on various components of foraging activity: comparison between European and Africanized honeybees in Brazil. *J. Apic. Res.* 18:110–115.

——— 1982. Honeybee foraging strategies at a food source in relation to its distance from the hive and the rate of sugar flow. *J. Apic. Res.* 21:139–150.

Nye, W. P., and O. Mackensen. 1968. Selective breeding of honeybees for alfalfa pollination: fifth generation and backcrosses. *J. Apic. Res.* 7:21–27.

——— 1970. Selective breeding of honeybees for alfalfa pollen collection: with tests in high and low alfalfa pollen collection regions. *J. Apic. Res.* 9:61–64.

Oertel, E. 1956. Observations on the flight of drone honeybees. *Ann. Entomol. Soc. Amer.* 49:497–500.

Oettingen-Spielberg, T. 1949. Über das Wesen der Suchbienen. *Z. vergl. Physiol.* 31:454–489.

Ohtani, T. 1974. Behavior repertoire of adult drone honeybee within observation hives. *J. Fac. Sci. Hokkaido Univ.* (ser. 6) 19:706–721.

Ohtani, T., and H. Fukuda. 1977. Factors governing the spatial distribution of adult drone honeybees in the hive. *J. Apic. Res.* 16:14–26.

Olaerts, E. 1956. Der Stoffwechsel der Bienen in Beziehung zu deren Aktivität und Temperatur. *Ber. Int. Bienenz. Kongr. Wien.* 16:70.

Onions, G. W. 1912. South African "fertile-worker bees." *Agric. J. Union S. Africa* 3:720–728.

——— 1914. South African "fertile worker bees." *Agric. J. Union S. Africa* 7: 44–46.

Ortiz-Picon, J. M., and L. Diaz-Flores. 1972. Etude structurale, optique, et electronique, des glandes hypopharyngiennes de *Apis mellifera. Consejo Superior Investig. Cient. Trabajos* 64:223–240.

Osanai, M., and H. Rembold. 1968. Entwicklungsabhängige mitochondriale Enzymaktivitäten bei den Kasten der Honigbiene. *Biochem. Biophys. Acta* 162:22–31.

Otis, G. W. 1980. The swarming biology and population dynamics of the Africanized honey bee. Ph.D. diss., Univ. of Kansas.

——— 1982. Weights of worker honeybees in swarms. *J. Apic. Res.* 21:88–92.

Otis, G. W., M. L. Winston, and O. R. Taylor, Jr. 1981. Engorgement and dispersal of Africanized honeybee swarms. *J. Apic. Res.* 20:3–12.

Owen, M. D. 1978a. Venom replenishment, as indicated by histamine, in honey bee (*Apis mellifera*) venom. *J. Insect Physiol.* 24:433–437.

——— 1978b. Histamine in the venom of queen and worker bees (*Apis mellifera* L.). In *Toxins: animal, plant and microbial,* ed. P. Rosenberg. Oxford and New York, Pergamon Press.

Owens, C. D. 1971. The thermology of wintering honey bee colonies. *Tech. Bull., U.S. Dep. Agric.* 1429:1–32.

Page, R. E. 1980. The evolution of multiple mating behavior by honey bee queens (*Apis mellifera* L.). *Genetics* 96:263–273.

——— 1981. Protandrous reproduction in honey bees. *Envir. Entomol.* 10:359–361.

——— 1986. Sperm utilization in social insects. *Ann. Rev. Entomol.* 31:297–320.

Page, R. E., and E. H. Erickson. 1984. Selective rearing of queens by worker honey bees: kin or nestmate recognition? *Ann. Entomol. Soc. Amer.* 77: 578–580.

Page, R. E., and R. A. Metcalf. 1982. Multiple mating, sperm utilization, and social evolution. *Amer. Nat.* 119:263–281.

——— 1984. A population investment sex ratio for the honey bee (*Apis mellifera* L.). *Amer. Nat.* 124:680–702.

Page, R. E., R. B. Kimsey, and H. H. Laidlaw. 1984. Migration and dispersal of spermatozoa in spermathecae of queen honeybees (*Apis mellifera* L.). *Experientia* 40:182–184.

Pain, J. 1961. Sur la phéromone des reines d'abeilles et ses effets physiologiques. *Ann. Abeille* 4:73–153.

——— 1973. Pheromones and hymenoptera. *Bee World* 54:11–24.

Pain, J., and M. Barbier. 1981. The pheromone of the queen honeybee. *Naturwissenschaften* 68:429–430.

Pain, J., and J. Maugenet. 1966. Recherches biochimiques et physiologiques sur le pollen emmagsiné par les abeilles. *Ann. Abeille* 9:209–236.

Pain, J., and M. M. B. Roger. 1978. Rythme circadien des acides céto-9-décène-oïque, phéromone de la reine, et hydroxy-10 décène-2-oïque des ouvrières d'abeilles *Apis mellifica ligustica* S. *Apidologie* 9:263–272.

Pain, J., M. Barbier, and M. M. B. Roger. 1967. Dosages individuels des acides céto-9-décène-2-oïque et hydroxy-10-décène-2-oïque dans les têtes des reines et des ouvrières d'abeilles. *Ann. Abeille* 10:45–52.

Pain, J., M. M. B. Roger, and J. Theurkauff. 1972. Sur l'existence d'un cycle annuel de la production de phéromone (acide céto-9 décène-2-oïque) chez les reines d'Abeilles (*Apis mellifera ligustica* Spinola) *C. R. Acad. Sci.* 275:2399–2402.

——— 1974. Mise en évidence d'un cycle saisonnier de la teneur en acides céto-9 et hydroxy-9 décène-2-oïque des têtes de reines vierges d'abeille. *Apidologie* 5:319–355.

Park, O. W. 1922. Time and labor factors involved in gathering nectar and pollen. *Amer. Bee J.* 62:254–255.

——— 1923a. Flight studies of the honeybee. *Amer. Bee J.* 63:71.

——— 1923b. Water stored by bees. *Amer. Bee J.* 63:348–349.

——— 1923c. Behavior of water carriers. *Amer. Bee J.* 63:553.

——— 1925a. The minimum flying weight of the honeybee. Report of the Iowa State Apiarist, pp. 83–90.

——— 1925b. The storing and ripening of honey by honeybees. *J. Econ. Entomol.* 18:405–410.

——— 1927. Studies on the evaporation of nectar. *J. Econ. Entomol.* 20:510–516.

——— 1928a. Further studies of the evaporation of nectar. *J. Econ. Entomol.* 21:882–887.

——— 1928b. Time factors in relation to the acquisition of food by the honeybee. *Res. Bull. Iowa Agric. Exp. Sta.* 108:183–225.

Parker, R. L. 1926. The collection and utilization of pollen by the honeybee. *Mem. Cornell Agric. Exp. Sta.* 98:1–55.

Patel, N. G., M. H. Haydak, T. A. Gochnauer. 1960. Electrophoretic components of the proteins in honeybee larval food. *Nature* 186:633–634.

Peer, D. F. 1957. Further studies on the mating range of the honey bee *Apis mellifera* L. *Can. Entomol.* 89:108–110.

Percival, K. K. 1954. Wings in the wild. *Brit. Bee J.* 82:28–29, 34, 82.

Perepelova, L. 1928a. Laying workers, the ovipositing of the queen, and swarming. *Bee World* 10:69–71.

——— 1928b. Materials concerning the biology of the bee: the work of the bees in the hive. *Opuit Pas.* 11:492–502 (in Russian).

——— 1928c. Nurse bees. *Opuit Pas.* 12:551–557 (in Russian).

——— 1947. Ways of increasing breeding in colonies. *Pchelovodstvo* 4:10–14 (in Russian).

Petit, I. 1963. Etude in vitro de la croissance des larves d'abeilles. *Ann. Abeille* 6:35–52.

Petrunkewitsch, A. 1901. Die Richtungskorper und ihr Schicksal im befruchteten und unbefruchteten Bienenei. *Zool. Jb. Anat.* 14:573.

Pflumm, W. 1969. Beziehungen zwischen Putzverhalten und Sammelbereitschaft bei der Honigbiene. *Z. vergl. Physiol.* 64:1–36.

Pflumm, W., and K. Wilhelm. 1982. Olfactory feedback in the scent marking behaviour of foraging honeybees at the food source. *Physiol. Entomol.* 7:203–207.

Pflumm, W., C. Peschke, K. Wilhelm, and H. Cruse. 1978. Einfluss der in einem flugraum Kontrollierten-Trachtverhältnisse auf das Duftmarkieren und die Abflugmagenfüllung der Sammelbiene. *Apidologie* 9:349–362.

Pham, M. H., B. Roger, and J. Pain. 1982. Variation en fonction de l'âge des ouvrières d'abeilles (*Apis mellifica ligustica* S.) du pouvoir d'attraction d'un extrait de phéromones royales. *Apidologie* 13:143–155.

Phillips, E. F., and G. S. Demuth. 1914. The temperature of the honeybee cluster in winter. *Bull. U.S. Dep. Agric.* 93:1–16.

Pickett, J. A., I. H. Williams, and A. P. Martin. 1982. (Z)-11-Eicosen-1-ol, an important new pheromonal component from the sting of the honey bee *Apis mellifera* L. (Hymenoptera, Apidae). *J. Chem. Ecol.* 8:163–176.

Pickett, J. A., I. H. Williams, A. P. Martin, and M. C. Smith. 1980. Nasonov pheromone of the honeybee *Apis mellifera* L. (Hymenoptera: Apidae), I. Chemical characterization. *J. Chem. Ecol.* 6:425–434.

Pickett, J. A., I. H. Williams, M. C. Smith, and A. P. Martin. 1981. Nasonov pheromone of the honeybee, *Apis mellifera* L. (Hymenoptera, Apidae), III. Regulation of pheromone composition and production. *J. Chem. Ecol.* 7:543–554.

Portillo, J. del. 1936. Beziehungen zwischen den Öffnungswinkeln der Ommatidien, Krümmung und Gestalt der Insektenaugen und ihrer funktionellen Aufgabe. *Z. vergl. Physiol.* 23:100–145.

Portugal Araujo, V. de. 1971. The Central African bee in South America. *Bee World* 52:116–121.

Punnett, E. N., and M. L. Winston. 1983. Events following queen removal in colonies of European-derived honey bee races. *Insectes Sociaux* 30:376–383.

Queeny, E. M. 1952. The wandorobo and the honey guide. *Nat. Hist.* 61:392–396.

Raven, P. H., and D. I. Axelrod. 1974. Angiosperm biogeography and past continental movements. *Ann. Mo. Bot. Gard.* 61:539–673.

Renner, M. 1960. Das Duftorgan der Honigbiene und die physiologische Bedeutung ihres Lockstoffes. *Z. vergl. Physiol.* 43:411–468.

Renner, M., and M. Baumann. 1964. Über Komplexe von subepidermalen Drüsenzellen (Duftdrüsen?) der Bienenkönigin. *Naturwissenschaften* 51: 68–69.

Renner, M., and G. Vierling. 1977. Die Rolle des Taschendrusenpheromons biem Hochzeitsflug der Bienenkonigin. *Behav. Ecol. Sociobiol.* 2:329–338.

Ribbands, C. R. 1949. The foraging method of individual honey-bees. *J. Anim. Ecol.* 18:47–66.

——— 1952. Division of labour in the honeybee community. *Proc. Roy. Soc. London* (B) 140:32–42.

——— 1953. *The behaviour and social life of honeybees.* London, Bee Research Association. (Repr., 1964, New York, Dover.)

——— 1954. The defense of the honeybee community. *Proc. Roy. Soc. London* (B) 142:514–524.

——— 1955. The scent perception of the honeybee. *Proc. Roy. Soc. London* (B) 143:367–379.

Riches, H. R. C. 1982. Hypersensitivity to bee venom. *Bee World* 63:7–22.

Rinderer, T. E. 1981. Volatiles from empty comb increase hoarding by the honey bee. *Anim. Behav.* 29:1275–76.

Rinderer, T. E., A. M. Collins, and K. W. Tucker. 1985. Honey production and underlying nectar harvesting activities of Africanized and European honeybees. *J. Apic. Res.* 24:161–167.

Rinderer, T. E., A. B. Bolten, A. M. Collins, and J. R. Harbo. 1984. Nectar-foraging characteristics of Africanized and European honeybees in the neotropics. *J. Apic. Res.* 23:70–79.

Roberts, W. C., and S. Taber III. 1965. Egg-weight variance in honey bees. *Ann. Entomol. Soc. Amer.* 58:303–306.

Robinson, F. A., and E. Oertel. 1975. Sources of nectar and pollen. In *The hive and the honeybee,* ed. C. P. Dadant, pp. 283–302. Hamilton, Ill., Dadant.

Robinson, G. E. 1982. A unique beekeeping enterprise in Colombia. *Bee World* 63:43–46.

——— 1984. Worker and queen honey bee behavior during foreign queen introduction. *Insectes Sociaux* 31:254–263.

——— 1985. Effects of a juvenile hormone analogue on honey bee foraging behaviour and alarm pheromone production. *J. Insect Physiol.* 31:277–282.

Robinson, G. E., B. A. Underwood, and C. E. Henderson. 1984. A highly specialized water-collecting honey bee. *Apidologie* 15:355–358.

Rösch, G. A. 1925. Untersuchungen über die Arbeitsteilung im Bienenstaat,

I. Teil: Die tätigkeiten im normalen Bienenstaate und ihre Beziehungen zum Alter der Arbeitsbienen. *Z. vergl. Physiol.* 2:571–631.

——— 1927. Über die Bautatigkeit im Bienenvolk und das Alter der Baubienen. *Z. vergl. Physiol.* 6:264–298.

——— 1930. Untersuchungen über die Arbeitsteilung im Bienenstaat, II. Teil: Die Tätigkeiten der Arbeitsbienen unter experimentell veränderten bedingunen. *Z. vergl. Physiol.* 12:1–71.

Rosov, S. A. 1944. Food consumption by bees. *Bee World* 25:94–95.

Roth, M. 1965. La production de chaleur chez *Apis mellifica. Ann. Abeille* 8:5–77.

Rothenbuhler, W. C. 1964. Behavioral genetics of nest cleaning in honey bees, IV. Responses of F2 and backcross generations to disease-killed brood. *Amer. Zool.* 4:111–123.

Rothenbuhler, W. C., J. M. Kulincevic, and V. C. Thompson. 1979. Successful selection of honeybees for fast and slow hoarding of sugar syrup in the laboratory. *J. Apic. Res.* 18:272–278.

Ruttner, F. 1956a. Zur Frage der spermaübertragung bei der Bienenkönigin. *Insectes Sociaux* 3:351–359.

——— 1956b. The mating of the honey bee. *Bee World* 3:2–15, 23–24.

——— 1962. Drohnensammelplätze. *Bienenvater* 83:45–47.

——— 1966. The life and flight activity of drones. *Bee World* 47:93–100.

——— 1968. Les races des abeilles. In *Traité de biologie de l'abeille,* vol. 1., ed. R. Chauvin, pp. 27–44. Paris, Masson.

——— 1975a. Races of bees. In *The hive and the honeybee,* ed. C. P. Dadant, pp. 19–38. Hamilton, Ill., Dadant.

——— 1975b. African races of honeybees. *Proc. XXV Inter. Apic. Congress, Grenoble,* pp. 325–344.

——— 1977. The problem of the Cape bee (*Apis mellifera capensis* Escholtz): parthenogenesis, size of population, evolution. *Apidologie* 8:281–294.

Ruttner, F., and B. Hesse. 1981. Rassenspezifische Unterschiede in Ovarentwicklung und Eiablage von weisellosen Arbeiterinnen der Honigbiene *Apis mellifera* L. *Apidologie* 12:159–183.

Ruttner, F., and V. Maul. 1983. Experimental analysis of reproductive interspecies isolation of *Apis mellifera* L. and *Apis cerana* Fabr. *Apidologie* 14: 309–327.

Ruttner, F., and H. Ruttner. 1965. Untersuchungen über die Flugaktivität und das Paarungsverhalten der Drohnen, II. Beobachtungen an Drohnensammelplätzen. *Z. Bienenforsch.* 8:1–9.

——— 1966. Untersuchungen über die Flugaktivität und das Paarungsverhalten der Drohnen, III. *Z. Bienenforsch.* 8:332–354.

Ruttner, F., N. Koeniger, and H. J. Veith. 1976. Queen substance bei eierlegenden Arbeiterinnen der Honigbiene (*Apis mellifera* L.). *Naturwissenschaften* 63:434–435.

Ruttner, F., D. Pourasghar, and D. Kauhausen. 1985. Die honigbienen des Iran, II. *Apis mellifera meda* Skorikow, die persische biene. *Apidologie* 16:241–264.

Ruttner, F., L. Tassencourt, and J. Louveaux. 1978. Biometrical-statistical analysis of the geographic variability of *Apis mellifera* L. *Apidologie* 9:363–381.

Ruttner, H. 1976. Untersuchungen über die Flugaktivität und das Paarungsverhalten der drohnen. *Apidologie* 7:331–341.

Ruttner, H., and F. Ruttner. 1972. Untersuchungen über die Flugaktivität und das Paarungsverhalten der Drohnen, V. Drohnensammelplatze und Paarungsdistanz. *Apidologie* 3:203–232.

Rutz, W., L. Gerig, H. Wille, and M. Lüscher. 1976. The function of juvenile hormone in adult worker honeybees, *Apis mellifera. J. Insect Physiol.* 22:1485–1491.

Rutz, W., H. Imboden, E. R. Jaycox, H. Willie, L. Gerig, and M. Luscher. 1977. Juvenile hormone and polyethism in adult worker honeybees (*Apis mellifera*). Proceedings of the Eighth Congress of the International Union for Study of Social Insects, pp. 26–27.

Sakagami, S. F. 1953. Untersuchungen über die Arbeitsteilung in einem Zwergvolk der Honigbiene: Beiträge zue Biologie des Bienenvolkes, *Apis mellifera* L. *I. Jap. J. Zool.* 11:117–185.

——— 1954. Occurrence of an aggressive behavior in queenless hives, with considerations of the social organization of honeybee. *Insectes Sociaux* 4:331–343.

——— 1958. The false queen: fourth adjustive response in dequeened honeybee colonies. *Behaviour* 13:280–296.

Sakagami, S. F., and H. Fukuda. 1968. Life tables for worker honeybees. *Res. Pop. Ecol.* 10:127–139.

Sakagami, S. E., T. Matsumura, and K. Ito. 1980. *Apis laboriosa* in Himalaya, the little known world largest honeybee (Hymenoptera, Apidae). *Insecta Matsumurana* 19:47–77.

Sammataro, D. and A. Avitabile. 1978. *The beekeeper's handbook.* New York, Charles Scribner's Sons.

Schifferer, G. 1952. *Uber die Entfernungsangabe bei den Tänzen der Bienen* (lehramtsarbeit). Munich, Naturwissenschaft Fakultät, Universität München.

Schmid-Hempel, P. 1984. The importance of handling time for the flight directionality in bees. *Behav. Ecol. Sociobiol.* 15:303–309.

Schmid-Hempel, P., A. Kacelnik, and A. I. Houston. 1985. Honeybees maximize efficiency by not filling their crop. *Behav. Ecol. Sociobiol.* 17:61–66.

Schmidt, J. O. 1982. Biochemistry of insect venoms. *Ann. Rev. Entomol.* 27:339–368.

Schneider, S. S., J. A. Stamps, and N. E. Gary. 1986a. The vibration dance of the honey bee, I. Communication regulating foraging on two time scales. *Anim. Behav.* 34:366–385.

——— 1986b. The vibration dance of the honey bee, II. The effects of foraging success on daily patterns of vibration activity. *Anim. Behav.* 34:386–391.

Schönitzer, K., and M. Renner. 1984. The function of the antenna cleaner of the honeybee (*Apis mellifica*). *Apidologie* 15:23–32.

Schricker, B. 1965. Die Orientierung der Honigbiene in der Dämmerung, zugleich ein Beitrag zur Frage der Ocellenfunktion bei Bienen. *Z. vergl. Physiol.* 49:420–458.

Schwan, B., and A. Martinovs. 1954. Studier över binas (*Apis mellifica*) pollendrag, Ultuna (Schweden). *Stat. Hus. Medd.* 57, 35 pp.

Schwartz, R. 1955. Über die Riechschärfe der Honigbiene. *Z. vergl. Physiol.* 37:180–210.

Scott, C. D. 1986. Biology and management of wild bee and domesticated honey bee pollinators for tree fruit pollination. Ph.D. diss., Simon Fraser Univ., Burnaby, B.C.

Seeley, T. D. 1974. Atmospheric carbon dioxide regulation in honey-bee (*Apis mellifera*) colonies. *J. Insect Physiol.* 20:2301–5.

——— 1977. Measurement of nest cavity volume by the honey bee (*Apis mellifera*). *Behav. Ecol. Sociobiol.* 2:201–227.

——— 1978. Life history strategy of the honey bee *Apis mellifera*. *Oecologia* 32:109–118.

——— 1979. Queen substance dispersal by messenger workers in honeybee colonies. *Behav. Ecol. Sociobiol.* 5:391–415.

——— 1982. Adaptive significance of the age polyethism schedule in honeybee colonies. *Behav. Ecol. Sociobiol.* 11:287–293.

——— 1983. Division of labor between scouts and recruits in honeybee foraging. *Behav. Ecol. Sociobiol.* 12:253–259.

——— 1985a. *Honeybee ecology.* Princeton, Princeton Univ. Press.

——— 1985b. The information-center strategy of honeybee foraging. *Fortsch. Zool.* 31:75–90.

Seeley, T. D., and R. D. Fell. 1981. Queen substance production in honey bee (*Apis mellifera*) colonies preparing to swarm. *J. Kans. Entomol. Soc.* 54:192–196.

Seeley, T. D., and B. Heinrich. 1981. Regulation of temperature in the nests of social insects. In *Insect thermoregulation,* ed. B. Heinrich, pp. 159–234. New York, Wiley.

Seeley, T. D., and R. A. Morse. 1976. The nest of the honey bee (*Apis mellifera*). *Insectes Sociaux* 23:495–512.

——— 1977. Dispersal behavior of honey bee swarms. *Psyche* 83:199–209.

——— 1978. Nest site selection by the honey bee *Apis mellifera*. *Insectes Sociaux* 25:323–337.

Seeley, T. D., and P. K. Visscher. 1985. Survival of honeybees in cold climates: the critical timing of colony growth and reproduction. *Ecol. Entomol.* 10:81–88.

Seeley, T. D., R. A. Morse, and P. K. Visscher. 1979. The natural history of the flight of honey bee swarms. *Psyche* 86:103–113.

Seeley, T. D., R. H. Seeley, and P. Akratanakul. 1982. Colony defense strategies of the honeybees in Thailand. *Ecol. Monog.* 52:43–63.

Sekiguchi, K., and S. F. Sakagami. 1966. Structure of foraging population and related problems in the honeybee, with considerations on the division of labor in bee colonies. *Hokkaido Nat. Agric. Exp. Sta. Rep.* 69:1–65.

Serian-Back, E. 1961. Vitamine—wichtige Faktoren in der Bienenernahrung. *Z. Bienenforsch.* 5:234–237.

Shearer, D. A., and R. Boch. 1965. 2-Heptanone in the mandibular gland secretion of the honey-bee. *Nature* 206:530.

Shearer, D. A., R. Boch, R. A. Morse, and F. M. Laigo. 1970. Occurrence of 9-oxodec-trans-2-enoic acid in queens of *Apis dorsata, Apis cerana,* and *Apis mellifera. J. Insect. Physiol.* 16:1437–41.

Shuel, R. W. 1975. The production of nectar. In *The hive and the honeybee,* ed. C. P. Dadant, pp. 265–282. Hamilton, Ill., Dadant.

Shuel, R. W., and S. E. Dixon. 1959. Studies in the mode of action of royal jelly in honeybee development. *Can. J. Zool.* 37:803–813.

——— 1960. The early establishment of dimorphism in the female honeybee (*Apis mellifera* L.). *Insectes Sociaux* 7:265–282.

——— 1973. Regulatory mechanisms in caste development in the honeybee, *Apis mellifera* L. Proceedings of the Seventh Congress of the International Union for Study of Social Insects, pp. 349–360.

Shuel, R. W., S. E. Dixon, and G. B. Kinoshita. 1978. Growth and development of honeybees in the laboratory on altered queen and worker diets. *J. Apic. Res.* 17:57–68.

Simpson, J. 1957. Observations on colonies of honey-bees subjected to treatments designed to induce swarming. *Proc. Roy. Entomol. Soc. London* (A) 32:185–192.

——— 1958. The factors which cause colonies of *Apis mellifera* to swarm. *Insectes Sociaux* 5:77–95.

——— 1959. Variation in the incidence of swarming among colonies of *Apis mellifera* throughout the summer. *Insectes Sociaux* 6:85–99.

——— 1960. The functions of the salivary glands of *Apis mellifera. J. Insect Physiol.* 4:107–121.

——— 1961. Nest climate regulation in honeybee colonies. *Science* 133:1327–33.

——— 1966. Repellency of the mandibular gland scent of worker honey bees. *Nature* 209:531–532.

——— 1973. Influence of hive-space restriction on the tendency of honeybee colonies to rear queens. *J. Apic. Res.* 12:183–186.

——— 1979. The existence and physical properties of pheromones by which worker honeybees recognize queens. *J. Apic. Res.* 18:233–249.

Simpson, J., and S. M. Cherry. 1969. Queen confinement, queen piping and swarming in *Apis mellifera* colonies. *Anim. Behav.* 17:271–278.

Simpson, J., and S. P. Greenwood. 1975. Results of restricting the brood space of honeybee colonies. *J. Apic. Res.* 14:51–55.

Simpson, J., and E. Moxley. 1971. The swarming behaviour of honeybee colonies kept in small hives and allowed to outgrow them. *J. Apic. Res.* 10:109–113.

Simpson, J., and I. B. M. Riedel. 1963. The factor that causes swarming by honeybee colonies in small hives. *J. Apic. Res.* 2:50–54.

Simpson, J., I. B. M. Riedel, and N. Wilding. 1968. Invertase in the hypopharyngeal glands of the honeybee. *J. Apic. Res.* 7:29–36.

Singh, S. 1950. Behavior studies of honeybees in gathering nectar and pollen. *Bull. Cornell Agric. Exp. Sta.* 288:1–59.

Slessor, K. N., G. G. S. King, D. R. Miller, M. L. Winston, and T. L. Cutforth. 1985. Determination of chirality of alcohol or latent alcohol semiochemicals in individual insects. *J. Chem. Ecol.* 11:1659–67.

Slifer, E. H., and S. S. Sekhon. 1961. Fine structure of the sense organs on the antennal flagellum of the honey bee *Apis mellifera* L. *J. Morphol.* 109:351–362.

Smith, F. G., 1959. A note on the capping activities of an individual honeybee. *Bee World* 40:153–154.

Smith, M. V. 1960. *Beekeeping in the tropics.* London, Longmans.

——— 1961. The races of honeybees in Africa. *Bee World* 42:255–260.

——— 1974. Relationship of age to brood-rearing activities of worker honey bees, *Apis mellifera* L. *Proc. Ent. Soc. Ont.* 105:128–137.

Snodgrass, R. E. 1956. *Anatomy of the honey bee.* Ithaca, Cornell Univ. Press.

Southwick, E. E., and J. N. Mugaas. 1971. A hypothetical homeotherm: the honeybee hive. *Comp. Biochem. Physiol.* 40a:935–944.

Spangler, H. G., and S. Taber III. 1980. Defensive behavior of honeybees toward ants. *Psyche* 77:184–189.

Standifer, L. N., and J. P. Mills. 1977. The effect of worker honey bee diet and age on the vitamin content of larval food. *Ann. Entomol. Soc. Amer.* 70:691–694.

Stanley, R. G., and H. F. Linskens. 1974. *Pollen: Biology, biochemistry, and management.* New York, Springer-Verlag.

Stauffer, P. H. 1979. A fossilized honeybee comb from late Cenozoic cave deposits at Batu Caves, Malay Peninsula. *J. Paleontol.* 53:1416–21.

Storer, T. I., and G. H. Vansell. 1935. Bee eating proclivities of the striped skunk. *J. Mammal.* 16:118–121.

Stort, A. C. 1974a. Genetic study of aggressiveness in two subspecies of *Apis mellifera* in Brazil, I. Some tests to measure aggressiveness. *J. Apic. Res.* 13:33–38.

——— 1974b. Genetical study of aggressiveness of two subspecies of *Apis mellifera* in Brazil, IV. Number of stings in the gloves of the observer. *Behav. Genetics* 5:269–274.

——— 1975a. Genetic study of aggressiveness of two subspecies of *Apis mellifera* in Brazil, II. Time at which the first sting reached a leather ball. *J. Apic. Res.* 14:171–175.

——— 1975b. Genetic study of the aggressiveness of two subspecies of *Apis mellifera* in Brazil, V. Number of stings in the leather ball. *J. Kansas Entomol. Soc.* 48:381–387.

Stort, A. C., and N. Barelli. 1981. Genetic study of olfactory structures in the antennae of two *Apis mellifera* subspecies. *J. Kansas Entomol. Soc.* 54:352–358.

Strang, G. E. 1970. A study of honey bee drone attraction in the mating response. *J. Econ. Entomol.* 63:641–645.

Taber, S., III. 1954. The frequency of multiple mating of queen honey bees. *J. Econ. Entomol.* 47:995–998.

——— 1958. Concerning the number of times queen bees mate. *J. Econ. Entomol.* 51:786–789.

——— 1964. Factors influencing the circadian flight rhythm of drone honey bees. *Ann. Entomol. Soc. Amer.* 57:769–775.

Taber, S., III, and C. D. Owens. 1970. Colony founding and initial nest design of honey bees, *Apis mellifera* L. *Anim. Behav.* 18:625–632.

Taber, S., III, and W. C. Roberts. 1963. Egg weight variability and its inheritance in the honey bee. *Ann. Entomol. Soc. Amer.* 56:473–476.

Taranov, G. F. 1947. Occurrence and development of the swarming instinct in the colony. *Pchelovodstvo* 2:44–54 (in Russian).

Taranov, G. F., and L. V. Ivanova. 1946. Observations on the behaviour of the queen in the colony. *Pchelovodstvo* 2(3):35–39 (in Russian).

Taylor, O. R. 1977. The past and possible future spread of Africanized honeybees in the Americas. *Bee World* 58:19–30.

——— 1984a. An aerial trap for collecting drone honeybees in congregation areas. *J. Apic. Res.* 23:18–20.

——— 1984b. A mating tube for studying attractiveness of queen honeybees and mating behaviour of drones. *J. Apic. Res.* 23:21–24.

Taylor, O. R., R. W. Kingsolver, and G. W. Otis, 1986. A neutral mating model for honey bees (*Apis mellifera* L.). *J. Apic. Res.* (in press).

Thompson, V. C., and W. C. Rothenbuhler. 1957. Resistance to American foulbrood in honeybees, II. *J. Econ. Entomol.* 50:731–737.

Tischer, J. 1940. Ueber die Kerkunft der gelben Farbstoffe des Bienenwachses. *Hoppe-Seyl. Z.* 267:14–22.

Todd, F. E., and C. B. Reed. 1970. Brood measurement as a valid index to the value of honey bees as pollinators. *J. Econ. Entomol.* 63:148–149.

Tribe, G. D., and D. J. C. Fletcher. 1977. Rate of development of the workers of *Apis mellifera adansonii* L. In *African bees: their taxonomy, biology, and economic use,* ed. D. J. C. Fletcher, pp. 115–119. Pretoria, Apimondia.

Trojan, E. 1930. Die Dufoursche Drüse bei *Apis mellifica. Z. Morphol. Tiere* 19:678–685.

Tucker, K. 1958. Automictic parthenogenesis in the honey bee. *Genetics* 43:299–316.

Tuenin, T. A. 1926. Concerning laying workers. *Bee World* 8:90–91.

Tulloch, A. P. 1980. Beeswax composition and analysis. *Bee World* 61:47–62.

Tushmalova, N. A. 1958. Formation of the food reflex in bees. *Pchelovodstvo* 35:25–29 (in Russian).

Vandenberg, J. D., D. R. Massie, H. Shimanuki, J. R. Peterson, and D. M. Poskevich. 1985. Survival, behavior, and comb construction by honey bees, *Apis mellifera,* in zero gravity aboard NASA shuttle mission STS-13. *Apidologie* 16:369–383.

Vansell, G. H. 1942. Factors affecting the usefulness of honeybees in pollination. *Circ. U.S. Dept. Agric.* 650:1–31.

Vansell, G. H., and C. S. Bisson. 1935. Origin of color in western beeswax. *J. Econ. Entomol.* 28:1001–2.

Vareschi, E. 1971. Duftunterscheidung bei der Honigbiene—Einzelzell-

Ableitungen und Verhaltensreaktionen. *Z. vergl. Physiol.* 75:143–173.

Velthuis, H. H. W. 1970. Queen substances from the abdomen of the honey bee queen. *Z. vergl. Physiol.* 70:210–222.

——— 1972. Observations on the transmission of queen substances in the honey bee colony by the attendants of the queen. *Behaviour* 41:105–129.

Velthuis, H. H. W., and J. van Es. 1964. Some functional aspects of the mandibular glands of the queen honeybee. *J. Apic. Res.* 3:11–16.

Velthuis, H. H. W., F. J. Verheijen, and A. J. Gottenbos. 1965. Laying worker honeybee: similarities to the queen. *Nature* 207:1314.

Verheijen-Voogd, C. 1959. How worker bees perceive the presence of their queen. *Z. vergl. Physiol.* 41:527–582.

Verma, S., and F. Ruttner. 1983. Cytological analysis of the thelytokous parthenogenesis in the Cape honeybee *Apis mellifera capensis* Escholtz. *Apidologie* 14:41–57.

Vierling, G., and M. Renner. 1977. Die Bedeutung des Sekretes der Tergittaschendrösen fur die Attraktivität der Bienenkönigin gegenuber jungen Arbaiterinnen. *Behav. Ecol. Sociobiol.* 2:185–200.

Visscher, P. K. 1983. The honey bee way of death: necrophoric behaviour in *Apis mellifera* colonies. *Anim. Behav.* 31:1070–76.

——— 1986. Kinship discrimination in queen rearing by honey bees (*Apis mellifera*). *Behav. Ecol. Sociobiol.* 18:453–460.

Visscher, P. K., and T. D. Seeley. 1982. Foraging strategy of honeybee colonies in a temperate deciduous forest. *Ecology* 63:1790–1801.

Vivino, A. E., and L. S. Palmer. 1944. The chemical composition and nutritional value of pollens collected by bees. *Arch. Insect Biochem.* 4:129–136.

Voogd, S. 1955. Inhibition of ovary development in worker bees by extraction fluid of the queen. *Experientia* 11:181–182.

Waddington, K. D. 1980. Flight patterns of foraging bees in relation to artificial flower density and distribution of nectar. *Oecologia* 44:199–204.

——— 1982. Honey bee foraging profitability and round dance correlates. *J. Comp. Physiol.* 148:297–301.

Wadey, H. J. 1948. Section de chauffe? *Bee World* 29:11.

Walker, E. P., F. Warnick, S. E. Hamlet, K. I. Lange, M. A. Davis, H. E. Uible, and P. F. Wright. 1975. *Mammals of the world.* Baltimore, Johns Hopkins Univ. Press.

Waller, G. D. 1970. Attracting honeybees to alfalfa with citral, geraniol, and anise. *J. Apic. Res.* 9:9–12.

Weaver, N. 1957. Effects of larval age on dimorphic differentiation of the female honeybee. *Ann. Entomol. Soc. Amer.* 50:283–294.

——— 1966. Physiology of caste determination. *Ann. Rev. Entomol.* 11:79–102.

Weaver, N., A. H. Alex, and F. L. Thomas. 1953. Pollination of Hubam clover by honeybees. *Prog. Rep. Tex. Agric. Sta.* 1559:1–3.

Weaver, N., C. C. Weaver, and J. H. Law. 1964. The attractiveness of citral to foraging honeybees. *Prog. Rep. Tex. Agric. Exp. Sta.* 2324:1–7.

Wedmore, E. B. 1942. *A manual of beekeeping.* London, Arnold.

Wehner, R. 1972. Dorsoventral asymmetry in the visual field of the honey bee, *Apis mellifica*. *J. Comp. Physiol.* 77:256–277.

Wehner, R., and S. Strasser. 1985. The POL area of the honey bee's eye: behavioural evidence. *Physiol. Entomol.* 10:337–349.

Weipple, T. 1928. Futterverbrauch und Arbeitsleistung eines Bienenvolkes im Laufe eines. *Jahres. Arch. Bienenk.* 9:70–79.

Weiss, K. 1967. Zur vergleichenden Gewichtsbestimmung von Bienenköniginnen. *Z. Bienenforsch.* 9:1–21.

——— 1984. Regulierung des Proteinhaushaltes im Bienenvolk (*Apis mellifica* L.) durch Brutkannibalismus. *Apidologie* 15:339–354.

Wenner, A. M. 1963. The flight speed of honeybees: a quantitative approach. *J. Apic. Res.* 2:25–32.

——— 1971. *The bee language controversy: An experience in science.* Boulder, Colo., Educational Programs Improvement Corp.

Wenner, A. M., and D. L. Johnson. 1967. Honey bees: do they use the direction and distance information provided by their dancers? *Science* 158:1076–77.

Wenner, A. M., P. H. Wells, and F. J. Rohlf. 1967. An analysis of the waggle dance and recruitment in honey bees. *Physiol. Zool.* 40:317–344.

Whitcomb, W., Jr. 1946. Feeding bees for comb production. *Glean. Bee Cult.* 74:198–202, 247.

Whitcomb, W., Jr., and H. F. Wilson. 1929. Mechanics of digestion of pollen by the adult honey bee and the relation of undigested parts to dysentery of bees. *Res. Bull. Wisc. Agric. Exp. Sta.* 92:1–27.

White, J. W., Jr. 1975. Composition of honey. In *Honey: a comprehensive survey,* ed. E. Crane, pp. 157–206. London, Heinemann.

White, J. W., Jr., and O. N. Rudyj. 1978. The protein content of honey. *J. Apic. Res.* 17:234–238.

White, J. W., Jr., M. H. Subers, and A. I. Schepartz. 1963. Identification of inhibine, the antibacterial factor in honey, as hydrogen peroxide, and its origin in a honey glucose oxidase system. *Biochem. Biophys. Acta* 73:57–70.

Wilde, J. de, and J. Beetsma. 1982. The physiology of caste development in social insects. *Adv. Insect Physiol.* 16:167–246.

Wilhelm, K. T., and W. W. Pflumm. 1983. Über den Einfluss blumenhafter Düfte auf des Duftmarkieren der Sammelbiene. *Apidologie* 14:183–190.

Williams, I. H., J. A. Pickett, and A. P. Martin. 1981. The Nasonov pheromone of the honeybee, *Apis mellifera* L. (Hymenoptera: Apidae), II. Bioassay of the components using foragers. *J. Chem. Ecol.* 7:225–237.

——— 1982. Nasonov pheromone of the honeybee *Apis mellifera* L. (Hymenoptera, Apidae), IV. Comparative electroantennogram responses. *J. Chem. Ecol.* 8:567–574.

Williams, J. L. 1978. Insects: lepidoptera (moths). In *Honey bee pests, predators, and diseases,* ed. R. A. Morse, pp. 105–127. Ithaca, Cornell Univ. Press.

Wilson, E. O. 1971. *The insect societies.* Cambridge, Mass., Harvard Univ. Press.

Wilson, H. F., and V. G. Milum. 1927. Winter protection for the honeybee colony. *Res. Bull. Wisc. Agric. Exp. Sta.* 75:1–47.

Winston, M. L. 1979a. The proboscis of the long tongued bees: a comparative study. *Univ. Kans. Sci. Bull.* 51:631–667.

——— 1979b. Events following queen removal in colonies of Africanized honeybees in South America. *Insectes Sociaux* 26:373–381.

——— 1979c. Intra-colony demography and reproductive rate of the Africanized honeybee in South America. *Behav. Ecol. Sociobiol.* 4:279–292.

——— 1979d. The potential impact of the Africanized honeybee on apiculture in Mexico and Central America. *Amer. Bee J.* 119:584–586, 642–645.

——— 1980. Swarming, afterswarming, and reproductive rate of unmanaged honeybee colonies (*Apis mellifera*). *Insectes Sociaux* 27:391–398.

Winston, M. L., and L. A. Fergusson. 1985. The effect of worker loss on temporal caste structure in colonies of the honey bee (*Apis mellifera* L.). *Can. J. Zool.* 63:777–780.

Winston, M. L., and S. J. Katz. 1981. Longevity of cross-fostered honey bee workers (*Apis mellifera*) of European and Africanized races. *Can. J. Zool.* 59:1571–1575.

——— 1982. Foraging differences between cross-fostered honeybee workers (*Apis mellifera*) of European and Africanized races. *Behav. Ecol. Soc.* 10: 125–129.

Winston, M. L., and C. D. Michener. 1977. Dual origin of highly social behavior among bees. *Proc. Natl. Acad. Sci.* (U.S.) 74:1134–37.

Winston, M. L., and G. W. Otis. 1978. Ages of bees in swarms and afterswarms of the Africanized honeybee. *J. Apic. Res.* 17:123–129.

Winston, M. L., and E. N. Punnett. 1982. Factors determining temporal division of labor in bees. *Can. J. Zool.* 60:2947–52.

Winston, M. L., and O. R. Taylor. 1980. Factors preceding queen rearing in the Africanized honeybee (*Apis mellifera*) in South America. *Insectes Sociaux* 27:289–304.

Winston, M. L., J. A. Dropkin, and O. R. Taylor. 1981. Demography and life history characteristics of two honey bee races (*Apis mellifera*). *Oecologia* 48:407–413.

Winston, M. L., S. R. Mitchell, and E. N. Punnett. 1985. Feasibility of package honey bee (Hymenoptera: Apidae) production in southwestern British Columbia, Canada. *J. Econ. Entomol.* 78:1037–41.

Winston, M. L., G. W. Otis, and O. R. Taylor. 1979. Absconding behaviour of the Africanized honeybee in South America. *J. Apic. Res.* 18:85–94.

Winston, M. L., O. R. Taylor, and G. W. Otis. 1983. Some differences between temperate European and tropical African and South American honeybees. *Bee World* 64:12–21.

Winston, M. L., K. N. Slessor, M. J. Smirle, and A. A. Kandil. 1982. The influence of a queen-produced substance, 9HDA, on swarm clustering behavior in the honeybee *Apis mellifera* L. *J. Chem. Ecol.* 8:1283–88.

Wirtz, P. 1973. Differentiation in the honey bee larva. *Meded. Landbhoogesch. Wageningen*, no. 73-75, 1–155.

Wirtz, P., and J. Beetsma. 1972. Induction of caste differentiation in the honey bee (*Apis mellifera* L.) by juvenile hormone. *Entomologia Exp. Appl.* 15:517–520.

Witherell, P. C. 1971. Duration of flight and of interflight time of drone honeybees, *Apis mellifera*. *Ann. Entomol. Soc. Amer.* 64:609–612.

——— 1972. Flight activity and natural mortality of normal and mutant drone honeybees. *J. Apic. Res.* 11:65–75.

Woyke, J. 1960. Natural and artificial insemination of queen honey bees. *Pszczelnicze Zeszyty Naukowe* 4:183–273 (in Polish). Summarized in *Bee World* 43:21–25.

——— 1962. The hatchability of lethal eggs in a 2 sex-allele fraternity of honeybees. *J. Apic. Res.* 1:6–13.

——— 1963. What happens to diploid drone larvae in a honeybee colony? *J. Apic. Res.* 2:73–75.

——— 1964. Causes of repeated mating flights by queen honeybees. *J. Apic. Res.* 3:17–23.

——— 1969. A method of rearing diploid drones in a honeybee colony. *J. Apic. Res.* 8:65–74.

——— 1971. Correlations between the age at which honeybee brood was grafted, characteristics of the resultant queens, and results of insemination. *J. Apic. Res.* 10:45–55.

——— 1973a. Experiences with *Apis mellifera adansonii* in Brazil and Poland. *Apiacta* 8:115–116.

——— 1973b. Reproductive organs of haploid and diploid drone honeybees. *J. Apic. Res.* 12:35–51.

——— 1976a. Population genetic studies on sex alleles in the honeybee using the example of the Kangaroo Island bee sanctuary. *J. Apic. Res.* 15:105–123.

——— 1976b. Brood-rearing efficiency and absconding in Indian honeybees. *J. Apic. Res.* 15:133–143.

——— 1977. Cannibalism and brood-rearing efficiency in the honey-bee. *J. Apic. Res.* 16:84–94.

——— 1983. Dynamics of entry of spermatozoa into the spermatheca of instrumentally inseminated queen honeybees. *J. Apic. Res.* 22:150–154.

Woyke, J., and F. Ruttner. 1958. An anatomical study of the mating process in the honeybee. *Bee World* 39:3–18.

Yanase, T., and M. Kataoka. 1963. The microstructure of the dorsal ocelli of the worker honeybee. *Zool. Mag.* (Tokyo) 72:48–52 (in Japanese, with an English summary).

Yu, S. J., F. A. Robinson, and J. L. Nation. 1984. Detoxication capacity in the honeybee, *Apis mellifera* L. *Pestic. Biochem. Physiol.* 22:360–368.

Zander, E. 1916. Die Ausbildung des Geschlechtes bei der Honigbiene (*Apis mellifica* L.). *Z. Angew. Entomol.* 3:1–20.

Zeuner, F. E., and F. J. Manning. 1976. A monograph on fossil bees (Hymenoptera: Apoidea). *Bull. Brit. Mus. Nat. Hist.* 27:1–268.
Zmarlicki, C., and R. A. Morse. 1963. Drone congregation areas. *J. Apic. Res.* 2:64–66.
Zolotov, V. V., and L. I. Frantsevich. 1973. Orientation of bees by the polarized light of a limited area of the sky. *J. Comp. Physiol.* 85:25–36.

Author Index

Subject Index